Springer Series in Wireless Technology

Series editor

Ramjee Prasad, Aalborg, Denmark

Springer Series in Wireless Technology explores the cutting edge of mobile telecommunications technologies. The series includes monographs and review volumes as well as textbooks for advanced and graduate students. The books in the series will be of interest also to professionals working in the telecommunications and computing industries. Under the guidance of its editor, Professor Ramjee Prasad of the Center for TeleInFrastruktur (CTIF), Aalborg University, the series will publish books of the highest quality and topical interest in wireless communications.

More information about this series at http://www.springer.com/series/14020

Lei Guan

FPGA-based Digital Convolution for Wireless Applications

Springer

Lei Guan
Nokia Bell Labs
Dublin
Ireland

ISSN 2365-4139 ISSN 2365-4147 (electronic)
Springer Series in Wireless Technology
ISBN 978-3-319-51999-9 ISBN 978-3-319-52000-1 (eBook)
DOI 10.1007/978-3-319-52000-1

Library of Congress Control Number: 2016962045

Printed on acid-free paper

This Springer imprint is published by Springer Nature
The registered company is Springer International Publishing AG
The registered company address is: Gewerbestrasse 11, 6330 Cham, Switzerland

To three most important people in my life
Rumei, Shiyu and Leona

Preface

Here at first, I would like to put down some motivations and general thoughts behind this book. During my past 10 years working period in mixed academic and industry environment, I had many technical discussions and leisure chats with global researchers and electronic engineers, most of them are active pioneers in universities, leading research organizations and international technology companies. I found an interesting fact that majority of them are either "thinkers" (question-raising persons) or "doers" (solution-providing persons), not both at the same time, in other words, the one with nice idea on a topic accompanying a bunch of mathematical equations is not 100% sure whether the idea is fully valid in the real physical world or the one who takes care of building prototype systems as daily basis does not have much visibility of new emerging ideas. This is not surprised, there is gap, a relatively huge one, between emerging technology topics and cutting-edge physical implementations, this can be considered as cross-domain activity, it happens either because the "thinkers" or the "doers" are not interested in the other direction or because at particular period of the career, they could not move on due to the lack of time or simply due to the absence of a proper/attractive guide book or some readings on that topic.

This book aims at building a bridge to reduce the gap between "thinking-person" and "doing-person" for researchers, engineers, and senior university students who are currently working or are just entering in the hardware wireless system signal processing discipline, a cross-discipline of Wireless Communications, Digital Signal Processing, and Field Programmable Gate Array (FPGA) design. FPGAs, as scalable digital processing platforms, have been widely and heavily used in all types of hardware equipments in the modern wireless mobile networks (like the third generation (3G) and the fourth generation (4G) wireless base stations and corresponding infrastructures) and in our opinion they will be more densely utilized in the emerging fifth generation (5G) wireless systems.

Psychologically, we do not learn new things in an exactly organized sequential way, for example, very few people mastered the alphabets in an exact A to Z manner, and we actually project those letters to our brain in a kind of guided random manner according to our own interests and pre-acquired experience. This is why quite a large percentage of our higher education systems have started trying to create customized learning plans for different students with different backgrounds. Similarly in the technology R&D organizations, there is a natural need for promoting research and creating new ideas by cross-discipline approaches. How to make the best use of multi-domain knowledge is not a simple discussion work, it is a piece of art with innovative thinking and solid actions behind. Moreover, validated by scientific research, the majority of people are visual learners, which means a self-explained graphic illustration in a book is much better to tell an attractive technical story than just reading 500 recondite words.

Therefore in this book, at first I would like to introduce the fundamentals of using FPGA for digital signal processing purpose (Chap. 2), and then I will use one of the most important functions (i.e., linear convolution) in the signal processing society as an example, to illustrate what is convolution, why we need that in the wireless systems, and how it is working in the real world (Chap. 3). I will illustrate (by many figures and drawings besides explanations) how to quickly transfer this signal processing idea to a robust working digital system implementation on the FPGAs. With easy-to-read psychotechnical analysis, hands-on procedures and graphic illustrations, the reader will quickly grab the basic technical skills. And then, in the latter part of the book, I will move to detailed advanced topics including FPGA-based nonlinear convolution (Chap. 4) and FPGA-based fast linear convolution (Chap. 5) that are tightly associated with many attractive wireless communication applications, particularly the digital predistortion (DPD) applications for high efficient modern wireless transceivers and spatial multiplexing applications for the forthcoming 5G massive MIMO wireless networks. At the end, I will illustrate how to quickly and properly evaluate the FPGA-based DSP functions in a software (SW) & hardware (HW) co-operated platform (Chap. 6), which has been successfully utilized in many applications.

And thanks for choosing this book along your professional road, hope you will enjoy the contents and learn something you need. If you have any suggestions, comments, or questions with regard to the book and the corresponding topics, please contact me at DR.LEI.GUAN@GMAIL.COM.

Contents

About the Author

Lei Guan obtained B.E. in communications engineering from Harbin Institute of Technology (HIT), China, in 2006. Since then he has been actively working as a wireless transceiver researcher and FPGA-DSP engineer both in academic and industry. He was awarded M.E. in telecommunication systems from HIT and Ph.D. in electronic engineering from University College Dublin (UCD), Ireland. After working on an industrial-related research project in Trinity College Dublin (TCD) for two years, he joined Nokia Bell Labs as a Member of Technical Staff in early 2013 working on the activities both in the Mobile Radio Research Domain and Innovation Incubation Domain. He has 10 years R&D experience in advanced digitally assisted wireless transceiver architecture and his main responsibilities are MATLAB based DSP algorithm development, FPGA-based digital system implementation and RF transceiver hardware prototyping. He co-authored over 20 international papers on the prestige journals and top conferences and has filed 5 patent applications.

During his early career days, he played with Altera FPGA devices, including Cyclone, Cyclone II, and Stratix II devices. Starting from 2008, he started working with Xilinx FPGAs until now, including Spartan-3, Virtex-4, Virtex-5, Virtex-6, Virtex-7, Kintex-7, Zynq SoC, and Kintex Ultrascale. He has extensive experience with Xilinx FPGA design, digital logic design and verification, DSP subsystem design and validation, board-level development and HW debugging, static timing analysis, synthesize, implementation, floorplanning, IP core design, some SoC design, using both Xilinx ISE and Vivado design suits.

Abbreviations

ACLR	Adjacent Channel Leakage Ratio
ADC	Analog to Digital Converter
AGC	Automatic Gain Control
ANN	Artificial Neural Network
ASIC	Application-Specific Integrated Circuit
BBE	Baseband Equivalent
BOM	Bill Of Materials
BPF	Band Pass Filter
BRAM	Block Random Access Memory
CA	Complex Addition
CFR	Crest Factor Reduction
CM	Complex Multiplication
CNN	Convolutional Neural Network
CORDIC	COrdinate Rotational DIgital Computer
CPRI	Common Public Radio Interface
CPU	Centre Processing Unit
CW	Continuous Wave
DAC	Digital to Analog Converter
DD-ANN	Direct Dynamic Artificial Neural Network
DDC	Digital Down Conversion
DDR	Dynamic Deviation Reduction
DDS	Direct Digital Synthesis
DFE	Digital Front End
DFT	Discrete Fourier Transform
DIF	Decimation in Frequency
DIT	Decimation in Time
DM	Data Multiplexing
DMA	Direct Memory Access
DPD	Digital Predistortion
DPRAM	Dual-Port Random Access Memory

DSP	Digital Signal Processing
DUC	Digital Up Conversion
DUT	Device Under Test
E-CM	Element-wise Complex Multiplication
EDA	Electronic Design Automation
EVM	Error Vector Magnitude
EQ	Equalization
FIR	Finite Impulse Response
FFT	Fast Fourier Transform
FLC	Fast Linear Convolution
FPGA	Field Programmable Gate Array
GPU	Graphic Processing Unit
GMP	Generalized Memory Polynomial
GPP	General Purpose Processor
GSM	Global System for Mobile
GUI	Graphic User Interface
HDL	Hardware Description Language
HLS	High Level Synthesis
HW	Hardware
IC	Integrated Circuit
IF	Intermediate Frequency
IFFT	Inverse Fast Fourier Transform
ILA	Integrated Logic Analyzer
IOT	Internet of Things
IP	Intellectual Property
LC	Linear Convolution
LiFi	Light Fidelity
LNA	Low Noise Amplifier
LO	Local Oscillator
LPF	Low Pass Filter
LS	Least Squares
LSAS	Large-Scale Antenna System
LSB	Least Significant Bit
LTE	Long-Term Evolution
LTI	Linear Time Invariant
LUT	Look Up Table
MIMO	Multiple Input Multiple Output
MISO	Multiple Input Single Output
MP	Memory Polynomial
MSE	Mean Square Error
MSB	Most Significant Bit
MSPS	Mega Samples Per Second
MUX	Multiplexer
NCO	Numerically Controlled Oscillator
NMSE	Normalized Mean Square Error

NRT	Non Real Time
OFDM	Orthogonal Frequency Division Multiplexing
OLA	OverLap Add
OLS	OverLap Save
PA	Power Amplifier
PC	Personal Computer
PCB	Print Circuit Board
PCI-E	Peripheral Component Interface Express
PLL	Phase Lock Loop
PSD	Power Spectral Density
R&D	Research and Development
PAPR	Peak to Average Power Ratio
RAM	Random access memory
RAN	Radio Access Network
RD-ANN	Recursive Dynamic Artificial Neural Network
RF	Radio Frequency
RRH	Remote Radio Head
RX	Receiver
RT	Real Time
RTL	Register Transfer Logic
SDR	Software Defined Radio
SIMO	Single Input Multiple Output
SIR	Signal to Interference Ratio
SISO	Single Input Single Output
SMFB	Spatial Multiplexing Filter Bank
SOC	System On Chip
SPI	Serial Peripheral Interface
SW	Software
TB	Test Bench
TDM	Time Division Multiplexing
TRX	Transceiver
TX	Transmitter
UART	Universal Asynchronous Receiver Transmitter
UMTS	Universal Mobile Telecommunications System
USB	Universal Serial Bus
VHDL	VHSIC Hardware Description Language
VPU	Vision Processing Unit
WCDMA	Wideband Code Division Multiple Access

Chapter 1
FPGA-based Convolutions for Wireless Communications

Abstract This chapter will open the topic of the book, starting from the introduction of the emerging trends in the wireless communications and short descriptions of the convolution and FPGA in the wireless applications at the high-level.

1.1 Emerging Trends in Wireless Communications

Wireless communications started over one hundred years ago when Marconi invented radio, and nowadays we are at the era where human race are getting seamlessly surrounded by various types of wireless communications. After almost two decades, Wi-Fi and mobile cellular have been dominating the wireless access network. As ITU (International Telecommunications Union) pointed out in the report [1], seven billion people (95% of the people on this planet) live in an area that is covered by a mobile-cellular network.

Believe it or not, human civilization has entered into an infinite digitization era: people have already started talking to the machines; machines communicate with other smart devices; cars share driving status and safety information with other neighbour cars on the road or the traffic supervising drones in the sky, and so on. After born, humans are wirelessly connected with parents and other surroundings in a natural way, while the modern electronics, emerging devices and wireless networks are expanding the human possibilities of technology. Going out of our thinking box, wireless connectivity will serve as invisible links to connect humans and machines at all times, enabling wide-range social interactions regardless the boundary of locations and the cultural differences.

Along the path of wireless network evolution, we observed some simple but interesting points: every concrete step that moving forward was driven by certain human social needs. Starting at early 1990s, 2G wireless network, like Global System for Mobile (GSM), enabled us communicate with another person a thousand miles away via warm human voice, rather than reading a cold telegram text without any feeling. Early in 2000s, as internet putting global people together, we grew a self-willingness to be part of this connected sharing world 24 hours per day

L. Guan, *FPGA-based Digital Convolution for Wireless Applications*,
Springer Series in Wireless Technology, DOI 10.1007/978-3-319-52000-1_1

even when we were in motion, thus there came 3G wireless network with wider bandwidth for data communications, like Universal Mobile Telecommunications System (UMTS). Now, we are at the second half of 2010s, the growing social needs to share information and experience anywhere anytime in any format (voices, pictures, videos, video-rich multi-media contents) push network operators upgrading the wireless network to the fourth generation (4G), like long term evolution advanced (LTE-advanced). Human needs for better experience of technologies will never stop growing, neither will the carrier wireless communications. Augment reality (AR), virtual reality (VR), 3D holographic video based remote experience sharing and reconstructing applications are emerging on ever-more powerful devices and pushing the corresponding wireless communications technologies to reach new heights. A full picture and forward looking vision of the future networks has been revealed from Bell Labs perspective in the book [2] and concretely demonstrated by the series of Future X Day activities [3].

Telecom standardization groups have already started the necessary study items and standardization work for the so-called fifth generation (5G) wireless communications. Different than the previous generations, this time the technology will be evolved to extend human possibilities leading us to further understand ourselves regarding deep social needs and new forms of human-to-human, machine-to-machine and human-to-machine interactions. We believe the emerging trends of wireless communications for boosting end user experience contain the following essential perspectives:

- All spectrum access: As the only physical resource for delivering wireless communication services, the frequency spectrum is generally labelled as one of the most valuable but limited resources. Enforced by different governments and telecommunications regulatory bodies, the spectrum was sliced into multiple frequency bands for different applications. However, telecom frontiers and practitioners (e.g., telecom infrastructure companies, wireless stand-ups, research institutes, and universities/colleges) have been working on the techniques using multiple frequency bands simultaneously to create a virtualized single wideband wireless tunnel. Though there are quite a few technical issues to be solved, we see the trend towards all spectrum access in the future regardless the nature of the spectrum bands, they could be licensed cellular band or unlicensed WiFi band, they could be conventional cellular bands (300 MHz to 3 GHz) or centimetre wave bands (3–30 GHz) or millimetre wave bands (30–300 GHz), or visible light bands (430–790 THz, e.g., Light Fidelity LiFi), and so on.
- All devices access: During the golden times for personal computer (PC) industry, we witnessed that millions of PCs were connected to the internet for working or entertaining purposes globally, however people had to be somewhere physically to get access to the internet. Inspired by almost two decades' accumulation in both industrial technologies and business models, a big reality is now under incubation internationally: billions of wireless end "devices" are connecting together either globally or locally. Those devices could

be super powerful handsets in your pockets, single function temperature sensors in your house, electrical cars in the motion, traffic supervising drones, disaster aiding flying mobile basestations, robot-type multimedia toys, millions of industrial automation control units, and so on. For short, wireless access has been and will be one of the most important features of the future inter-active smart devices, which are "required" to gain access to the network locally or globally, constantly or periodically, actively or passively, straightforwardly or intelligently.

- All software-defined organic network architecture: Starting from the 1st generation (1G) mobile network to the latest 4G wireless network, Telcos (telecommunications companies) have been actively behaving as dominated wireless service providers. However as the rising IT companies (e.g., Facebook, Google) are trying to revolutionise the conventional telecom networks, we do see a trend that our wireless network gradually moves towards an all software-defined architecture—based organic and intelligent wireless network, which can adaptively reconfigure itself by software "manipulation" only (in a micro scale locally or a large scale globally) for providing satisfactory user experience in different mobility scenarios such as low speed mobility scenario (e.g., street walking), high speed mobility scenario (e.g., train travelling), temporal content dense scenario (e.g., concert in a stadium), network traffic predictable dense urban scenario (e.g., business park during the day) and so on. This is not an easy evolution, nowadays people may complain the service provided by the mobile networks, but currently there are a very large number of telecommunications engineers behind doing manually network configurations and optimizations on the specific hardware equipments. Empowering the network with intelligent self-adaptive capability will lead a future organic wireless network.

1.2 Convolutions in Wireless Communications

Modern wireless communication systems have been evolved significantly and they are relatively complex signal processing systems comparing to the 1st generation cellular basestations using fewer signal processing blocks. Telecom researchers and engineers have carried out lots of R&D work to advance two essential topics in the wireless communications: (1) how to understand or characterize the invisible wireless transmission and (2) how to efficiently deliver or pass the information (signal) from one device to another device wirelessly. Convolution, as a powerful mathematic tool, plays a very important role in both of these two areas. The term "convolution" was originally derived from Medieval Latin word "convolvere", which means "roll together". In mathematics, convolution describes a behavior how two functions are interactively producing a third function. And in wireless communications, convolution can be generally considered as a digital signal processing approach to make interactions with the signals for certain purposes, like removing some unwanted properties of the signals (filtering function), reshaping the signals against the wireless channel fading (equalization function) and so on. And this book will focus the detailed knowledge about the convolution and its applications in the wireless communications with detailed explanations both in the theory and practical implementations.

1.3 FPGAs in Wireless Communications

The modern wireless communication systems have been advanced by many signal processing algorithms, which need to be running at certain hardware processing units. Field programmable gate array (FPGA) is a type of digital integrated circuit (IC) chip providing re-configurable parallel processing capability, which is desired in the flexible software-defined wireless networks. FPGAs have been largely utilized in the modern wireless communication systems handling the complex signal processing and data transportation tasks. In the next Chapter, we will introduce the essential contents around the FPGA in terms of its functionalities, capabilities and applications for digital signal processing (DSP) purposes.

References

1. ITU (2016) ICT facts and figures 2016
2. Weldon MK (2016) The future X network: a Bell Labs perspective. CRC Press, New York
3. Nokia Bell Labs website http://www.bell-labs.com

Chapter 2
FPGA and Digital Signal Processing

Abstract This chapter will introduce the essential information of field-programmable gate-array (FPGA) and FPGA-based digital signal processing at system level without getting into too much detailed hardware design and implementation issues. The contents of this chapter will cover the following three topics: the state-of-the-art FPGAs, FPGA-based DSP basics and FPGA-based DSP system design. We'd like provide a concise system level view and use this chapter as a "soft" appetizer prior the "hard" main dishes.

2.1 Introduction

Different than general purpose integrated circuit (IC), like microcontroller, an application specific integrated circuit (ASIC) is a type of IC that is customized for a particular application. Usually, we categorize the ASICs into three types: (1) full-custom ones, (2) standard cell-based or semi-custom ones and (3) programmable ones. Among the programmable ASICs, field programmable gate array (FPGA) is one of the most important and widely utilized. Simply speaking, FPGA chips are manufactured in a ready-to-use state, users can customize the FPGA chips with software-defined "programs" to achieve desired functions and re-program the chips later for carrying out other functions.

In 1985, Xilinx [1] co-founders Ross Freeman and Bernard Vonderschmitt invented the first commercial FPGA—XC2064, one year before, Altera [2] released the first industrial reprogrammable logic device—EP300. After thirty years growing, the top two FPGA vendors regarding market shares at the moment are Xilinx and Altera (an Intel company since end of 2015), but quite a few semiconductor companies have been very actively engaged in FPGA related activities and business, such as Lattice Semiconductor [3] and Microsemi [4] (was Actel).

L. Guan, *FPGA-based Digital Convolution for Wireless Applications*,
Springer Series in Wireless Technology, DOI 10.1007/978-3-319-52000-1_2

2.2 State-of-the-Art FPGAs

FPGAs were born in the 1980s when digital system design engineers were struggling with the continuously evolving requirements for larger digital system design tasks. After three decades, nowadays FPGAs have already become one of the most popular platforms for building digital systems in many applications across a large range of industries. Modern FPGAs are not only providing parallel operations but also capable of high speed processing and massive data throughput interfacing. It is quite amazing to see the role of FPGAs has been changing along the time.

2.2.1 From Glue Logic to Scalable Computing Platform

At the early stage, FPGAs were quite small regarding the available number of logics on the chip, resulting that they were mainly used as glue logic to attach two or more logical modules together. In other words, the early FPGAs were widely utilized as the interfaces towards a number of off-the-shelf ICs or self-contained electronic modules for building a larger-size and more complicated digital application. At that time, the main resources on the FPGAs were merely logic and memory. Thanks to the growth of semiconductor process technology in an evolution manner, the state-of-the-art FPGAs are fundamentally evolved in all dimensions, particularly the chip-architecture, logic density and on-chip hard-core functionalities.

Figure 2.1 illustrates the evolution of FPGA functionalities on a single chip from glue logic to the scalable computing platform with heterogeneous processing units including, hardware multipliers, soft-core CPU, ARM [5], graphic processing unit (GPU) [6], vision processing unit (VPU). Besides the those delicate processing units, quite a few other high-performance units can be equipped on a FPGA chip, such as high-speed serial transceivers (over 10 Gbps per lane), large on-chip random access memory (RAM), customizable external memory interfaces and many integrated interfaces, like SPI, USB, Ethernet, PCI-express and so on. Actually, we

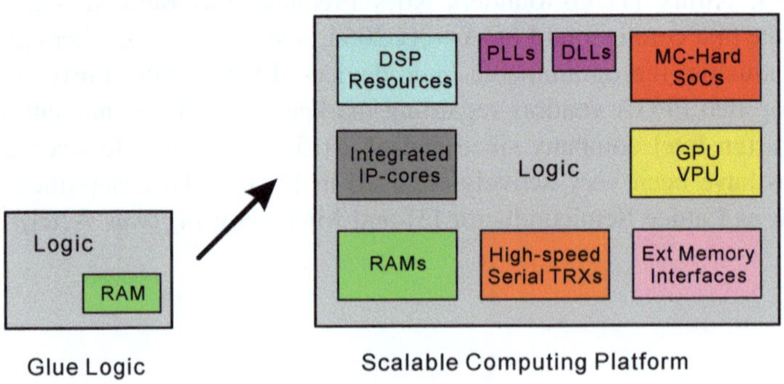

Fig. 2.1 FPGA evolution regarding functionalities on the chip

may give another name for this monster IC, like heterogeneous computing IC, since this chip architecture is far beyond the original FPGA concept.

2.2.2 From Control Logic to High-Speed Parallel Processing

FPGAs are inherently parallel processing capable ICs. This amazing feature enables very robust multi-branch simultaneous control functions associated to the multiple real-time co-operated machines. Without any surprise, FPGAs are quite popular and have been widely used in the industry as central control units monitoring and adjusting multiple devices/machines simultaneously.

As the analog switching performance of transistor has been improved continuously, the state-of-the-art FPGA fabrics can be reliably running at over 500 MHz processing. Together with its nature parallel operation, FPGAs not only provide super control logics for the applications with scaled number of devices, such as millions of Internet of things (IoT) terminals, but also enable the high-speed parallel data transmission and processing applications. Because of the above features, FPGAs have been heavily utilized in the telecom industry, e.g., on the communications infrastructures. Let's have a look at the modern remote radio head (RRH) in the wireless network infrastructures.

RRHs have become one the most important network equipments for both centralized and distributed wireless network architectures. In the forthcoming 5th generation (5G) wireless network, RRHs will be used as very important equipments to flexibly extend the coverage of centralized radio access network (C-RAN) or cloud-based radio access network (Cloud-RAN) with virtualized base-station function. The basic function of a RRH is to up-convert the digital baseband signal to the analog RF signal at the transmitter and down-convert the received analog RF signal back to digital baseband signal at the receiver branch. At one end, the RRHs are usually connected to the baseband processing units via CPRI interface using fibre as the real transmission media. At the other end, RRHs are positioned quite close to the antenna panel, so that short RF cables (saving very valuable RF power) can be used to connect them on the top of buildings. An example of simplified functional diagram of a typical wireless RRH is illustrated in Fig. 2.2. A typical RRH usually contains digital part and analog/RF part and FPGAs are normally used as the host at the digital part.

On the modern RRHs, like the ones deployed in the 4G wireless network, there are a few necessary functions are running in the FPGA chips working as digital front end (DFE) [7] including digital up-conversion (DUC), crest factor reduction (CFR), digital predistortion (DPD) and digital down-conversion (DDC). And some of them may have MIMO precoding module and MIMO equalization module to support multiple transceivers operations. Thanks to the state-of-the-art FPGA technology, we can see multiple similar function modules can be achieved on a

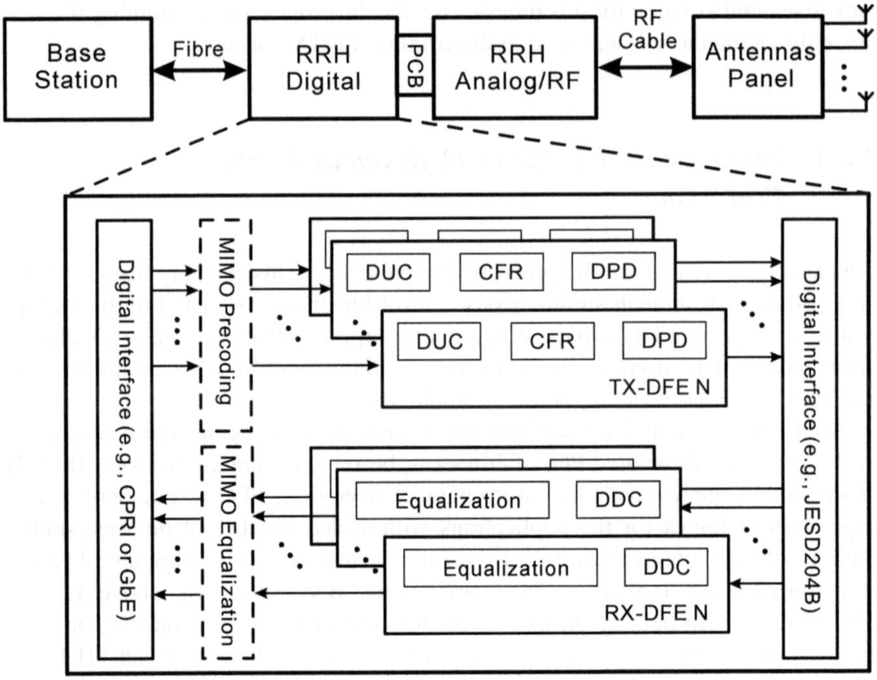

Fig. 2.2 Typical digital functions on the FPGA used on the RRH digital part

single chip for processing multiple data branches in parallel, saving quite a bit processing and interfaces power and reducing the size of print circuit board (PCB) and bill of materials (BOM) comparing to the multiple chips solution.

2.2.3 *From VHDL Design to Dense IP-Cores Integration*

The design philosophy of modern FPGA has changed. For simple glue logics, control functions or small-size digital applications, writing hardware description language (HDL), like VHDL, Verilog would be a quite efficient and powerful approach. However, as the chip design processing developing and the increasing number of transistors on the same silicon area, modern FPGAs are becoming monsters regarding the available number of logics and number of processing units. Building from scratch at register level won't be the optimal approach to do the FPGA design any more. Instead, pre-defined or customizable intellectual property (IP) core integration based design strategy will be more suitable for majority advanced DSP applications. IP-cores are reusable units of logic, cell or chip layout design provided by FPGA vendors, 3rd party companies or self-developed. By using IP-cores integration-based FPGA system design approach, we can significantly reduce the R&D cycle and minimize/eliminate the designer-in-the-loop risk (like buggy coding) that could be a very annoying issue.

In wireless infrastructures, digitally-assisted RF concept and corresponding techniques have been used to facilitate and improve the physical RF transmission. For example, CFR function can reduce the signal peak to average power ratio (PAPR) helping RF power amplifiers working in an efficient way; DPD function pre-distorts the signal to combat against the nonlinearities introduced by real RF power amplifiers. The emerging Massive MIMO wireless system [8] will utilize digital algorithms (precoding) to dynamically shape antenna beams minimizing the energy radiated towards the unwanted areas, so that the system capacity and signal to interference ratio (SIR) can be significantly improved.

For mature applications, like wireless infrastructure, since the standard functions have been well studied with pre-defined input/output module interfaces, we can easily build a complex digital system by directly integrating the commercial IP-cores from FPGA vendors or the ones from 3rd-Party IP-core vendors or the ones designed in-house.

For emerging applications of smart hardware innovation, like designing a drone, IP-core integration approach would be the first choice to rapidly create a ready-to-use drone with minimized risk at prototyping stage. A simplified functional diagram of a typical drone is shown in Fig. 2.3, where contains the essential hardware sub-systems including flight motor control sub-system, camera video processing sub-system, wireless transceiver sub-system and localization sub-system. On top of those sub-systems, there is a centre control function (contains four IP-cores taking care of the corresponding four sub-systems) that will initialize, re-configure, co-ordinate the sub-systems.

Moreover, using IP-core integration approach to design a digital system will provide smooth upgrading flexibility, for example, if we decide to use 3G signal,

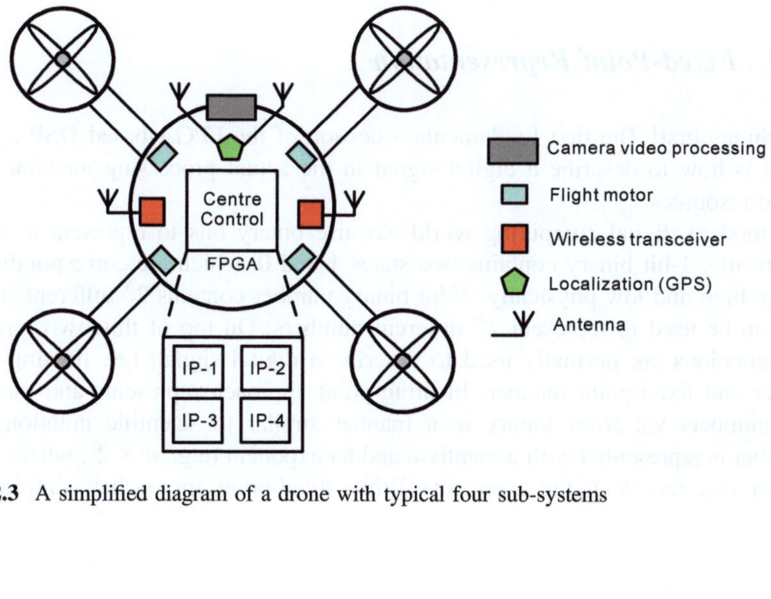

Fig. 2.3 A simplified diagram of a drone with typical four sub-systems

like wideband code division multiple access (WCDMA) type wireless air waveform to transmit the video information from the drone to the ground station, we could directly place a WCDMA baseband processing IP-core in the FPGA to handle the corresponding waveform-related processing. Then later, if the WCDMA type wireless signal cannot satisfy the increasing data rate requirement for high-definition video with high frame rate, we'd like to upgrade the air waveform to 4G orthogonal frequency-division multiplexing (OFDM) based, we only need to replace the WCDMA baseband processing IP-core with an OFDM baseband processing IP-core in the FPGA.

Above all, as modern digital systems are getting more complicated, it is quite difficult and not an economic way to start the design from zero. Simply, we are not going to invent the wheel again. IP-core integration should be the first choice for system architects and DSP engineers, since this approach significantly simplifies the FPGA design procedures, especially for large digital systems.

2.3 FPGA-based DSP Basics

Signal processing has been widely and heavily utilized to improve our daily life, any smart thing you can think of with even just a little bit intelligence will have certain signal processing units behind. Particularly comparing to the analog signal processing, the digital signal processing (DSP) has quite a few advantages including more flexibility, lower power consumption, higher reliability, higher accuracy and much better scalability. DSP functions can be achieved on different types of hardware processors, like microcontroller, ARM, CPU, GPU, SoC, FPGA and so on. In this chapter, we will focus on the essentials of FPGA-based digital signal processing approaches.

2.3.1 Fixed-Point Representation

First things first! The first fundamental question of the FPGA-based DSP system design is how to describe a digital signal in the actual processing machine with limited resources.

In modern digital computing world, we use binary bits to represent a signal. Specifically, 1-bit binary contains two states 1 and 0, which are corresponding to voltage high and low physically. N-bit binary number contains 2^N different stages, thus can be used to represent 2^N different numbers. On top of this, two types of representations are normally used to describe a digital signal, i.e., floating-point manner and fixed-point manner. Floating-point approach represents and manipulates numbers via N-bit binary in a manner similar to scientific notation, i.e., a number is represented with a mantissa and an exponent (e.g., $a \times 2^b$, where 'a' is the mantissa and 'b' is the exponent). While fixed-point approach is simpler and just use a fixed number of bits to express fractional numbers (e.g., $2^{N-M} \cdot 2^M$, where 'M' is the number of bits to express fractional numbers).

Modern FPGAs can handle both floating-point processing and fixed-point processing, however comparing to the fixed-point processing, floating-point processing will need significant resources and running power on FPGAs. Comparing to the fixed-point processing, floating-point processing generally provides higher accuracy with higher resolution at cost of higher resource utilization and higher power consumption. Luckily, proper selection of the fixed-point representation with dynamical tuning the fixed-point position can achieve satisfactory performance comparing to the floating-point processing with negligible differences for majority industrial applications. In short, with given tolerance, fixed-point processing would be the first choice for FPGA-based DSP system design.

For the sake of simplicity, we directly give an example of how to describe the signal by a 4-bit fixed-point representation with different number of fractional bits as shown in Table 2.1. The term Fix_N_N_F represents a N-bit signed signal with N_F-bit fractional part, while UFix_N_N_F represents a N-bit unsigned signal with N_F-bit fractional part. In signed representation like Fix_N_N_F, the most significant bit (MSB) will be used as sign, i.e., 0 stands for positive while 1 stands for negative, and 2's compliment representation is utilized (will describe in the next sub-section). Particularly, the Fix_N_0 and UFix_N_0 are corresponding to N-bit 2's complement binary and N-bit original binary, respectively.

Table 2.1 Comparison of 4-bit fixed-point binary numbers with different fractional bits

Binary				Interpreted values					
MSB to LSB				Fix_4_0	Fix_4_1	Fix_4_2	UFix_4_0	UFix_4_1	UFix_4_2
b_3	b_2	b_1	b_0	$b_3b_2b_1b_0.$	$b_3b_2b_1.b_0$	$b_3b_2.b_1b_0$	$b_3b_2b_1b_0.$	$b_3b_2b_1.b_0$	$b_3b_2.b_1b_0$
0	0	0	0	0	0.0	0.00	0	0.0	0.00
0	0	0	1	1	0.5	0.25	1	0.5	0.25
0	0	1	0	2	1.0	0.50	2	1.0	0.50
0	0	1	1	3	1.5	0.75	3	1.5	0.75
0	1	0	0	4	2.0	1.00	4	2.0	1.00
0	1	0	1	5	2.5	1.25	5	2.5	1.25
0	1	1	0	6	3.0	1.50	6	3.0	1.50
0	1	1	1	7	3.5	1.75	7	3.5	1.75
1	0	0	0	−8	−4.0	−2.00	8	4.0	2.00
1	0	0	1	−7	−3.5	−1.75	9	4.5	2.25
1	0	1	0	−6	−3.0	−1.50	10	5.0	2.50
1	0	1	1	−5	−2.5	−1.25	11	5.5	2.75
1	1	0	0	−4	−2.0	−1.00	12	6.0	3.00
1	1	0	1	−3	−1.5	−0.75	13	6.5	3.25
1	1	1	0	−2	−1.0	−0.50	14	7.0	3.50
1	1	1	1	−1	−0.5	−0.25	15	7.5	3.75

2.3.2 Two's Complement Binary Arithmetic Basic

The second question of the FPGA-based DSP approach is how to carry out the arithmetic using binary presentations, especially with minus sign.

One may think using the most significant bit (MSB) as sign and the rest of the bits to represent the amplitude as natural representation could do the work. However this approach will cause a systematic unstable situation when there is a positive zero and a negative zero (e.g., under this approach, $(0000)_2$ refers to positive zero while $(1000)_2$ will refer to negative zero).

2's complement approach solves the above issue and is one of the most important representations in the modern digital computing world. Let's take a closer look from 1's complement approach. Given an N-bit binary number x, the 1's complement of x is defined as an operation to invert every bit of the original binary number. For example, consider a binary number $(0101)_2$ in natural representation, then 1's complement representation of this number is $(1010)_2$. Straightforwardly, 1's complement representation of an N-bit unsigned binary number can be derived by subtracting the number from the maximum value $2^N - 1$.

The so-called 2's complement of an N-bit binary number is defined as the complement with respect to the value 2^N, and can be achieved simply by adding one to its 1's complement representation. By using this approach, the positive and negative zero dilemmas can be eliminated.

Furthermore, taking both unsigned and signed binary into consideration, the 1's complement and 2's complement operations of an N-bit binary number x can be described as Eqs. (2.1) and (2.2), respectively.

$$x_{1's} = 2^N - 1 - |x| \tag{2.1}$$

$$x_{2's} = x_{1's} + 1 = 2^N - |x| \tag{2.2}$$

The 2's complement representation makes fundamental arithmetic operations (like addition, subtraction, multiplication) for signed binary number quite simple, actually those operations are identical to those for unsigned binary numbers, which is really suitable for implementation of DSP functions on FPGAs. For example, subtraction can be done in the same way as addition by changing one operand sign using 2's complement representation. We provide a simple example as below, calculating "5 − 2" in binary representation.

Approach 1: 4-bit subtraction operation using natural binary representation:

$$(0101)_2 - (0010)_2 = (0011)_2 \tag{2.3}$$

Approach 2: 4-bit addition operation, "5 − 2" = "5 + (−2)". Using 2's complement representation, decimal value of (−2) will be $2^N - |-2| = 14$ and its corresponding binary representation is $(1110)_2$, then the subtraction can be done in the way of addition, producing the correct results (the overflow bit is removed without affecting the calculation result). 2's complement of a positive number is itself.

$$(0101)_{2's} + (1110)_{2's} = (0011)_{2's} = (0011)_2 \tag{2.4}$$

2.3.3 *Real Number and Complex Number*

The third question of the FPGA-based DSP approach is how to describe a complex signal and how to perform complex arithmetic.

Simple signals contain the information that can be characterized by only real numbers, for example, voltage signal has amplitude information together with positive or negative information to indicate the direction of the voltage (positive or negative, e.g., +12 or −3.3 V). However some applications need complex representation of the signals. In wireless communications, a simple signal modulation scheme will shape a sine wave to encode information. For example, Eq. (2.5) illustrates a modulated signal sine wave $s(t)$ with amplitude A_c, carrier frequency f_c and phase information θ_c.

$$s(t) = A_c \cdot \cos(2\pi f_c t + \theta_c) \tag{2.5}$$

The polar representation of sine wave is shown in Fig. 2.4, where θ(t) represents the instantaneous phase information.

However, fast high-precise phase varying of a high-speed carrier sine wave will be very difficult to achieve in the hardware circuits, resulting in very expensive to build such a modulation system in the hardware. An alternative way is to use so-called IQ modulation scheme. Let's reform the equation Eq. (2.5) as below

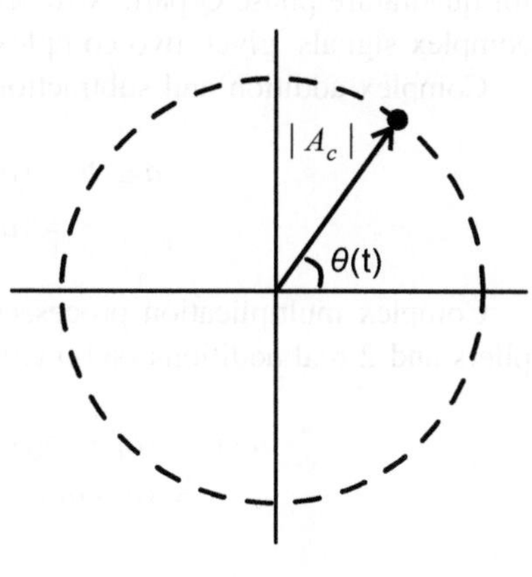

Fig. 2.4 A modulated sine wave signal in the polar coordinates

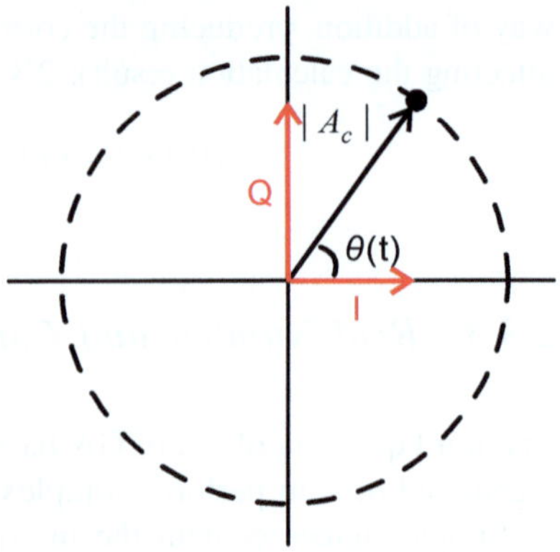

Fig. 2.5 A modulated sine wave signal in the Cartesian coordinate with I and Q data

$$\begin{aligned} A_c \cos(2\pi f_c t + \theta_c) &= A_c \cos(2\pi f_c t)\cos(\theta_c) - A_c \sin(2\pi f_c t)\sin(\theta_c) \\ &= I\ \cos(2\pi f_c t) - Q\ \sin(2\pi f_c t) \end{aligned} \tag{2.6}$$

where $I = \cos(\theta_c)$ represents the amplitude of in-phase carrier and $Q = \sin(\theta_c)$ represents the amplitude of quadrature-phase carrier. In this way, we transform the information from polar data system to the so-called Cartesian I and Q complex data system as shown in Fig. 2.5.

And the Eq. (2.5) can be simplified as below:

$$\begin{aligned} & s(t) = s_I(t) + s_Q(t) \\ & where \begin{cases} s_I(t) = I\ \cos(2\pi f_c t) \\ s_Q(t) = -Q\ \sin(2\pi f_c t) \end{cases} \end{aligned} \tag{2.7}$$

In the modern wireless systems, complex IQ signals can be simply represented by two signed binary numbers in the hardware, one for in-phase I part and other one for quadrature-phase Q part. Now let's have a closer look at the basic arithmetic of complex signals, given two complex numbers $a = a_I + ja_Q$ and $b = b_I + jb_Q$.

Complex addition and subtraction can be processed separately as below:

$$\begin{aligned} a \pm b &= (a_I + ja_Q) \pm (b_I + jb_Q) \\ &= (a_I \pm b_I) + j(a_Q \pm b_Q) \end{aligned} \tag{2.8}$$

Complex multiplication processed by the direct architecture with 4 real multipliers and 2 real additions (subtraction is considered as addition):

$$\begin{aligned} a \cdot b &= (a_I + ja_Q) \cdot (b_I + jb_Q) \\ &= (a_I \cdot b_I - a_Q \cdot b_Q) + j(a_I \cdot b_Q + a_Q \cdot b_I) \end{aligned} \tag{2.9}$$

Complex multiplication processed by the compact architecture with 3 real multipliers and 5 real additions:

$$
\begin{aligned}
a \cdot b &= (a_I + ja_Q) \cdot (b_I + jb_Q) \\
&= (a_I \cdot b_Q - a_Q \cdot b_Q) + j[(a_I + a_Q) \cdot (b_I + b_Q) - a_I b_I - a_Q b_Q]
\end{aligned} \tag{2.10}
$$

Please note that in the above examples, the addition operation between I part and Q part are just in place for illustration purpose, and I and Q signals will be two separated signals physically in the hardware.

2.3.4 Data Sampling Rate and Processing Rate

The fourth question of FPGA-based DSP approach is how to decide the data sampling rate and processing rate.

Based on the information theory, particular Nyquist-Shannon sampling theorem, in order to fully capture an analog signal of finite bandwidth, we need to sample the signal at a rate at least of the two times of signal bandwidth. In FPGA-based DSP system, the data sampling rate refers to the input digital signal refreshing rate which is associated with the type of signals. For example, in the 4G wireless network, one component carrier of long term evolution (LTE) advanced signal will need a corresponding 30.72 mega samples per second (MSPS) sampling rate. Each sample is an N-bit (e.g., 16-bit) complex binary number for both I-part and Q-part. In this case, the refreshing rate CLK_{DATA} of the input data will be 30.72 MHz and 32-bit complex IQ data will be updated per refreshing. Processing rate refers to the actual FPGA running clock CLK_{FPGA}, i.e., how fast the FPGA chip is operating for a given DSP task. Figure 2.6 illustrates the data rate and processing rate, where $CLK_{FPGA} = 6\ CLK_{DATA}$ as an example. Though the contents and refreshing rate of Data 1 and Data 2 are the same, the valid time of each sample are different in the two cases. Each sample of Data 2 is valid for one cycle of CLK_{DATA} while each sample of Data 1 is valid for one cycle of CLK_{FPGA}.

In general, the processing rate of FPGA CLK_{FPGA} should be at least as the same as input data sampling rate CLK_{DATA}. As the modern semiconductor technology improving, the state-of-the-art FPGAs can be reliably operating at couple of hundred MHz. Without any power constraints, the higher of the processor the better of DSP performance, simply because we can do more tasks within the same time period. Using higher-than required processing rate in the FPGA-based DSP-design has two direct appealing impacts: (1) lower processing latency and (2) lower processing resource utilization. Two examples are provided below.

An example of lower processing latency: given the data sampling rate $CLK_{DATA} = 4$ MHz and we are going to perform complex multiplication by a

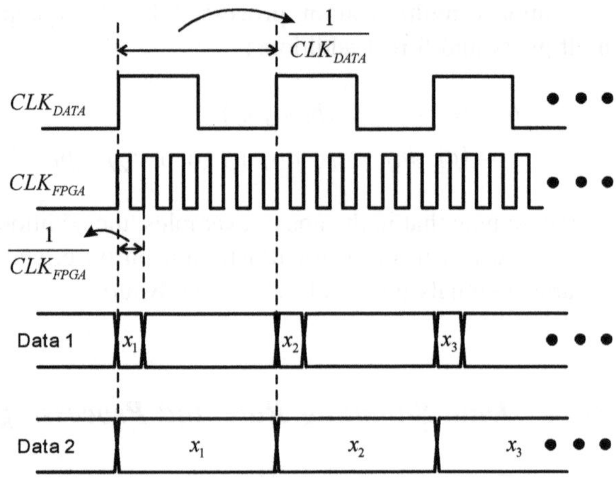

Fig. 2.6 Illustration of data rate and processing rate

complex multiplier IP-core with 4 clock processing delays. If the processing rate $CLK_{FPGA} = 4$ MHz, which is the same as the data sampling rate, the latency between sending complex data in until pushing complex data out is 4 clocks, which is $4 \times 1/4$ MHz = 1 ms latency. However, if the processing rate $CLK_{FPGA} = 40$ MHz, which is 10 times of data sampling rate, the latency will be $4 \times 1/40$ MHz = 0.1 ms in time, in other words, the processing latency has been reduced 10 times.

An example of lower processing resource utilization: given the data sampling rate $CLK_{DATA} = 4$ MHz and we are going to perform complex multiplication for 10 data branches simultaneously. If the processing rate $CLK_{FPGA} = 4$ MHz, which is the same as the data sampling rate, we will need 10 complex multiplier IP-cores operating in parallel with total 4 clock processing delay, which is 1 ms as the same as in the last example. However, if the processing rate $CLK_{FPGA} = 40$ MHz, which is 10 times of data sampling rate. Processing one complex multiplication will consume 4 clock cycles, which is 0.1 ms at the $CLK_{FPGA} = 40$ MHz, then within the same latency 1 ms, we can finish the complex multiplications for 10 branches using only one complex multiplier IP. We will explain more in the design strategy sub-section regarding this point.

2.3.5 Accuracy and Complexity

Without any optimization algorithms or design tricks, in general the complexity and accuracy of the FPGA-based DSP system blocks are a pair of trade-offs. Accuracy will be preliminary determined by the bit-width, i.e., enlarging the bit-width of a signal will directly increase the accuracy of the processing, but simultaneously resulting in the increments of the complexity (occupying more silicon area) due to the requirement for processing more bits per second.

The practical rules are quite simple: for performance-driven applications, we will use more resource to ensure high accurate processing, while for resource-constrained applications, we should maximize the processing accuracy by using the available resources. As design resources are in general constrained, in order to achieve better revenue to cost ratio for a product/solution, low complexity algorithms (achieving the same functionality) with smart design tricks that can minimize resources utilization will be always welcome in the DSP system design.

2.3.6 Dynamic Processing Range and Overflow

In a DSP system containing multiple processing stages, each stage may need a different processing range depending on the processing requirements. Figure 2.7 illustrates a generalized multiple stages processing system. Each DSP stage may increase the bit-width of the signals to combat against the so-called overflow issue due to the nature of binary processing. For example, a Fix_N_1_$N_{F,1}$ number multiplying by a Fix_N_2_$N_{F,2}$ number will result in a number in the form of Fix_$(N_1 + N_2)$_$(N_{F,1} + N_{F,2})$ to ensure the output is not out of "legal" representation range causing overflow problems. However, in a multi-stage DSP system, the bit-width may grow to a large number that is not possible or very difficult to handle by the next processing stage. Proper "re-arranging" functions are required.

For fixed-point processing, a dynamic range tuning function can be used to scale (left shift or right shift operation) and select (by slicing) the necessary/proper bit-widths for the next stage. Please note, those dynamic range tuning functions will cost nothing in the hardware.

Let's have a look at a quick example with meaningful settings. Assume a real multiplication operation, $a \times b = c$, where both inputs (a and b) are within the range of $[-1, 1)$ thus 16-bit fixed-point representation of inputs will be Fix_16_15. Without any range tuning function, the output c will be in the form of Fix_32_30 as full precision processing. Mathematically, if we carry out a multiplication between two numbers of range $[-1, 1)$, the output will be within the same range $[-1, 1)$. Therefore, if 16-bit resolution would be enough for the next stage processing, we can select the most left 16-bit out of 32-bit and move corresponding binary point to the 14-bit with regards to the least bit, e.g., Fix_16_14, this is equivalent to perform the quantization operation on the high dynamic range digital signals. In this way,

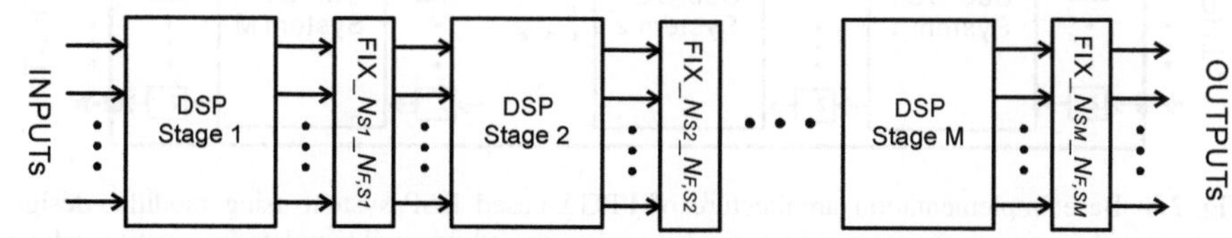

Fig. 2.7 Generalized M-stage signal processing with dynamic processing range

next stage will have 16-bit inputs as starting point. Best practice regarding this point is to scale the inputs and outputs of each processing stage to $[-1, 1)$, and tailor the representation by the available processing bit-width.

2.4 FPGA-based DSP System Design

Design a DSP system on the FPGA platform has been made quite easy and straightforward by the revolutionary efforts of FPGA vendors like Xilinx and Altera (An Intel Company). In the following sub-sections, we summarized the essential design strategies that will be helpful during FPGA-based DSP system design procedures. We recommend using those strategies from every beginning of a FPGA-based DSP system design project, eliminating the avoidable issues.

2.4.1 *Self-contained Modular Design Strategy*

Great wall was built by small pieces! A complex FPGA-based DSP system will be built by a few logical sub-functions, each of which is taking care of some processing tasks. Creating reasonable self-contained sub-modules (proper partitioning the FPGA regarding the DSP sub-systems) is the essential point to achieve a high performance DSP system on the FPGA platforms. The fundamental rule is to use so-called modular design strategy. If you have ever purchased some furniture from IKEA, you may notice that a lot of products contain some similar parts, so yes, they follow the modular design strategy. The parts they designed can be reused on different products, saving the cost and making the re-construction work easier. This self-contained modular design strategy together with re-timing registers between multiple sub-modules (as shown in the Fig. 2.8) is highly recommended as the base architecture of the FPGA-based DSP system (more detailed examples will be provided in following chapters), because:

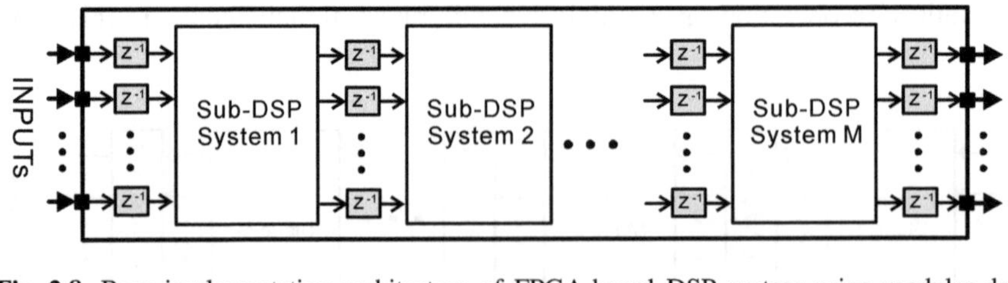

Fig. 2.8 Base implementation architecture of FPGA-based DSP system using modular design strategy with re-timing registers, which can be used as an architectural template for creating robust FPGA-based DSP systems

(1) Smaller sub-modules are relatively easy to handle both in the simulation and on silicon debugging/verification;
(2) Reasonable hierarchies of the sub-modules with pipelined registers can prevent FPGA timing issue for some certain degree and leave comfortable optimization space for the FPGA synthesize and implementation tool to properly route the logics in the chip. Also in this way, it is relatively easy to find the critical path that is associated with potential or existing timing hazards;
(3) Upgrading the modular designed DSP system is straightforward, i.e., simply replacing the sub-module by a modified one. This approach won't affect any other validated sub-modules and avoids the effort for re-designing the similar function blocks in different projects.

2.4.2 Time-Space Trade-off Implementation Strategy

Time-space trade-off implementation strategy naturally comes along with modern FPGA because of its inherent parallel processing capability. You may hear it before in the form of latency-resource trade-off on the CPU processing platform. The basic concept of this time-space trade-off enables very flexible design approaches: (1) in a resource-rich scenario (e.g., plenty of design resources on the chip), digital functions can be designed in the full parallel way to minimize the input to output latency, i.e., using more silicon processing area (space) to gain less processing latency (time); (2) in a resource-non-rich environment (e.g., limited resources assigned for achieving one digital function), the same physical resources have to be re-used to handle multiple tasks causing extra processing latency, i.e., reducing the resource utilization (space) by using multiple cycles processing (time). Figure 2.9 illustrates an example of time-space trade-off for multiple independent multiplication operations. Approach 1, i.e., full parallel architecture (when processing rate is equal to data rate) utilizes 4 multipliers to carry out 4 independent multiplication operations within one unit of time *t*, while the approach 2, i.e., full-serial architecture (when processing rate is equal to data rate) utilize 1 multiplier to process 4 multiplication operations but requiring 4 units of time.

As the silicon process improving, modern FPGAs can be running at high-speed, e.g., over 800 MHz on Xilinx Ultrascale FPGA device fabric has been reported, thus we can apply a technique, which is termed as time division multiplexing (TDM), to reuse the same physical resources in different time periods. In Fig. 2.9, the TDM idea is illustrated by approach 3, which contains the same full-serial architecture but in this case the processing rate is higher than data rate. Parallel inputs at data rate are multiplexed at processing rate and generating a high-speed serial data stream. In this way, 1 multiplier can process 4 multiplications within 4 shorter clock cycles (due to the higher processing rate), which is equivalent to one unit of time *t*. Comparing approach 3 and approach 1, it is not difficult to realize that

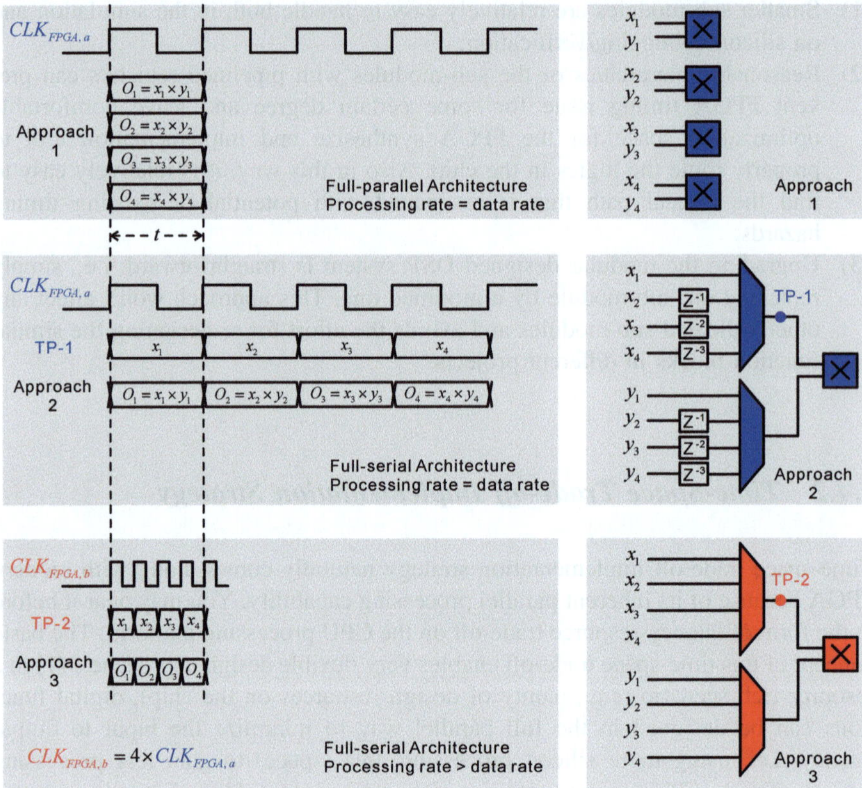

Fig. 2.9 Time-space trade-off illustration using an example of four multiplication operations

increasing the processing rate within physically achievable range will bring us the benefits of applying TDM approach, which will carry out the same DSP function with lower physical resource utilization.

2.4.3 Model-based Design Flow Using System Generator

Next, let's have a look how we can build a DSP system on a FPGA. First of all, DSP systems are naturally structured in steps, in other words, DSP systems can be abstracted/defined by a few mathematical equations. This is the beauty of the DSP systems, i.e., so-called black-box effect. Given the inputs and some pre-defined system parameters, the black-boxes (DSP systems) produce the desired outputs if the "boxes" have been properly described. For the sake of simplicity, we name those boxes as models, which reflect certain functions or some particular input-to-output mapping relationship using mathematical equations.

Quite a few design approaches are available to create a DSP system on the FPGA, the top three of most popular approaches include:

(1) Hardware description language (HDL)-based approach, e.g., using VHDL, Verilog, System Verilog to directly describe the hardware behavior at the RTL level;
(2) High level synthesis (HLS)-based approach, e.g., using C type high level language synthesis tool to describe the function behavior and create corresponding digital hardware logics;
(3) Model-based approach, e.g., using and extending MATLAB/Simulink [9] blocks to enable hardware design with high-level abstractions.

Among those three, because of its easy-to-use and quick-to-verify features at the system level, the model-based approach is quite suitable for DSP design on FPGAs. Xilinx provides a high-performance DSP design tool, i.e., System Generator for DSP [10], to help DSP/FPGA Engineers create production-quality DSP functions on a FPGA with short R&D time. We will use this Xilinx tool to tell the story in this book.

Based on Xilinx recommendations and our FPGA-based DSP design experience since 2006, System Generator can be quite helpful at the R&D prototyping stage for many scenarios. At early R&D stage of a brand new product, your algorithms are under developing so you may want to explore the algorithms regarding its system-level performance and corresponding complexity but without translating those “half-finished” algorithms into real hardware. Or at the prototyping stage for adding in new features on top of the existing products, you only need to design an add-on DSP function as part of the existing whole system design, however you have physical resource constraints so you may want to evaluate the new add-on DSP function first regarding its complexity with regards to the available physical resources left. In common, the model-based design flow using System Generator contains the following procedures (Table 2.2).

We will use this design flow as the base procedures in Chaps. 3–5.

2.4.4 *Overview of MATLAB and Simulink in FPGA DSP Design*

Due to its powerful scientific computing capability and flexible functional blocks re-customizability, MATLAB together with Simulink, has become the preferred algorithm R&D tool that is widely utilized in the both academia and industry. Precisely, MATLAB and Simulink are quite helpful in FPGA-based DSP system design in the following three perspectives:

(1) DSP Design preparation: Inherently organized in a column vector way, MATLAB provides quite efficient mathematical computing for vectors-based signal processing, especially in the wireless communication area. Before

Table 2.2 Model-based design flow using system generator

Step	Task	Target
1	Reference algorithm exploration using floating-point Simulink modules and partition the big design into smaller functional sub-modules	Ensure algorithm with desired functionalities and proper sub-modules partitioning
2	System parameters exploration regarding the interfaces between neighbour sub-modules	Ensure only necessary information pass to the next processing stage
3	Design constraints analysis with regards to the actual FPGA resources and latency requirement	Be clear with design constrains, both available resource and latency
4	Choose base implementation architecture, target device and corresponding clock parameters	Ensure proper system setting, e.g., data rate and processing rate …
5	Design the first sub-module using Xilinx tool-box in the System Generator; and verify the designed sub-module by comparing its outputs with counterparts in the reference model given the same inputs as excitation	Ensure the functionality and the accuracy of the designed sub-module meet the requirement
6	Repeat step 5 for the rest sub-modules	Ensure the functionalities of the sub-modules and interfaces
7	Whole system verification from original inputs to the final outputs	Ensure the functionality and accuracy of the whole module meet the requirement
8	Perform hardware-in-loop co-simulation (if possible)	Ensure the designed DSP system is functionally working on the real FPGA device
9	Generate design outputs, which can be VHDL, Verilog or synthesizable netlist, …	Produce the design files for integration

starting anything in the hardware, it is critical to ensure that the "ideas", i.e., DSP algorithms, are functional with given assumptions. We call this algorithm simulation. MATLAB makes this simulation relatively easy, and moreover, with the help of Simulink, you will be able to see the dynamic processing flow for your algorithms. In order words, MATLAB and Simulink provide a chance to know the inside out performance and functionality of your algorithms.

(2) DSP Design: Once you have a good and reliable algorithm simulation, translating the DSP algorithm into FPGA hardware can be easily achieved by MATLAB and Simulink together with Xilinx System Generator. We will provide more detailed examples from Chaps. 3–5.

(3) DSP Design validation: In order to make your DSP system work properly in the hardware, you will need to verify the functionality and timing performance of your design. For functionality validation, we recommend carry out comparison between designed module and reference MATLAB/Simulink model step by step. For timing validation, we recommend avoiding timing issue at

the early design stage using modular design strategy with re-timing registers, rather than solving the timing issue at the implementation stage. Simply because, the early you solve this timing problem, the less messy of your DSP system design project, and the shorter time you spend on the whole project and the more robust of your DSP module.

2.5 Conclusions

In this chapter, we quickly re-flashed the FPGA evolution and FPGA-based digital signal processing at the system level. And moreover, we introduced the essential and fundamental elements in the FPGA-based DSP system design including signal representation and model-based DSP system design strategy, which will be utilized in the following chapters with real examples in the wireless applications, especially the digital convolution applications.

References

1. Xilinx company website. http://www.xilinx.com
2. Altera (now part of Intel) company website. http://www.altera.com
3. Lattice Semiconductor company website. http://www.latticesemi.com
4. Microsemi company website. http://www.microsemi.com
5. ARM company website. http://www.arm.com
6. Nvidia company website, http://www.nvidia.com
7. Luo F (2011) Digital front-end in wireless communications and broadcasting. Cambridge University Press, Cambridge
8. Weldon MK (2016) The future X network: a Bell Labs perspective. CRC Press, New York
9. MathWorks company website. http://www.mathworks.com
10. Xilinx (2015) Model based DSP design using system generator. User Guide, UG897

Chapter 3
FPGA-based Linear Convolution

Abstract This chapter will focus on the essentials of discrete linear convolution and how it actual works on the FPGA platforms. The contents include the linear convolution basics, varies FPGA implementation architectures in details and typical applications of linear convolution in wireless communications, such as digital up conversion, digital down conversion, equalization filters and poly-phase filters.

3.1 Introduction

Convolution is a very efficient mathematical method used by the scientists and engineers to analyze and process the signals, like extracting some information from the signals or removing some information from the signals or adding some extra information to the signals. It is such a powerful analysis and processing tool that it has been widely utilized in many applications including communications signal processing, image processing, machine vision, deep learning for artificial intelligent (AI), natural language processing and so on. In this book, we will focus on the convolution applications for the telecommunications, particularly the real-time applications in the telecom transceivers.

3.2 Linear Convolution Basics

As technologies evolving, telecommunication equipments, e.g., transceivers, become very complex that comprise of a few self-contained sub-systems taking care of different static or dynamic processing or transportation tasks. From system analysis point view, if a system whose output does not depend on time, it belongs to a class of systems, which are named as time invariant system. As shown in Fig. 3.1, a system is characterized by a so-called transfer function, and input signal $x(t)$ passing transfer function $f\,[\cdot]$ results in output signal $y(t)$.

L. Guan, *FPGA-based Digital Convolution for Wireless Applications*,
Springer Series in Wireless Technology, DOI 10.1007/978-3-319-52000-1_3

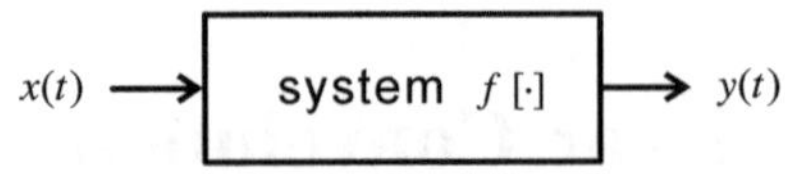

Fig. 3.1 An abstract system with transfer function $f\,[\cdot]$, input signal $x(t)$ and output signal $y(t)$

If the system $f\,[\cdot]$ is time invariant, then given any time shifted input, i.e., $x(t + t_o)$, outputs will be the time shifted version of $y(t)$, i.e., $y(t + t_o)$. The following equation describes the behavior of time invariant system $f\,[\cdot]$.

$$\begin{cases} y(t) = f[x(t)] \\ y(t+t_o) = f[x(t+t_o)] \end{cases} \tag{3.1}$$

On top of time invariant property, the so-called linear time invariant (LTI) system has an unique feature, i.e., given K inputs $x_k(t)$ (k = 1, 2, …, K) and system transfer function $f\,[\cdot]$, if

$$y_k(t) = f[x_k(t)] \tag{3.2}$$

Then,

$$\sum_k g_k y_k(t+t_o) = \sum_k g_k f[x_k(t+t_o)] \tag{3.3}$$

where $g_1, g_2, \ldots, g_k$ are real numbers.

The beauty of the LTI system is that, we can use single function named as impulse response to entirely describe an LTI system. And the outputs of a continuous-time LTI system are only determined by the linear convolution of the input to the system impulse response, i.e., given an input $x(t)$ and system impulse response $h(t)$, the output can be derived according to the Eq. 3.4 as below

$$y(t) = x(t) * h(t) \tag{3.4}$$

where the symbol * denotes the linear convolution operation.

This LTI behavior is not only true for continuous-time system but also valid for the discrete-time system. For the sake of simplicity in the FPGA-based digital signal processing context, we will use discrete representations in the rest of the book, e.g., using $x(n)$, $h(n)$ and $y(n)$ to replace $x(t)$, $h(t)$ and $y(t)$, respectively. And we will not distinguish discrete-time convolution from continuous-time convolution, unless there is a need to separate them for clarification purpose. Figure 3.2 illustrates a discrete-time LTI system with impulse response $h(n)$.

Fig. 3.2 A discrete LTI system with system impulse response $h(n)$

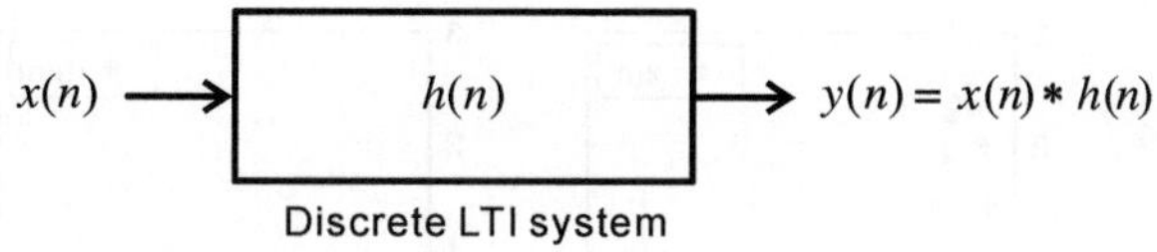

3.2.1 Time Domain Perspective

From time domain point view, linear convolution describes a processing how to derive the system output by given inputs and system impulse response. Specifically, the system output is the aggregation of the weighted input samples and its neighbour weighted samples in the time domain. Mathematically, linear convolution can be represented as below,

$$y(n) = x(n) * h(n) = \sum_{m=-\infty}^{\infty} x(m)h(n-m) = \sum_{m=-\infty}^{\infty} h(m)x(n-m) \tag{3.5}$$

In the definition, the impulse response $h(n)$ is infinite in the time domain for generalization case, however in the most of the practical and stable digital signal processing systems, the impulse response is finite. For example, a linear convolution with an M-tap finite impulse response can be represented as,

$$y(n) = x(n) * h(n) = \sum_{m=0}^{M-1} h(m)x(n-m) \tag{3.6}$$

Sometimes for simplicity, we present the linear convolution in a compact vector format as below,

$$\begin{aligned} y(n) &= \mathbf{H} \cdot \mathbf{X}^T \\ \textit{where} \quad \mathbf{H} &= [h(0),\ h(1),\ \ldots, h(M-1)] \\ \mathbf{X} &= [x(n),\ x(n-1),\ \ldots,\ x(n-(M-1))] \end{aligned} \tag{3.7}$$

In the time domain, the linear convolution with finite impulse response (FIR) describes the memory behavior of a system, e.g., given a LTI system with 4-tap impulse response vector $\mathbf{H} = [h(0), h(1), h(2), h(3)]$, the system output is only associated with the current and past 3 inputs samples as below,

$$y(n) = x(n) \cdot h(0) + x(n-1) \cdot h(1) + x(n-2) \cdot h(2) + x(n-3) \cdot h(3) \tag{3.8}$$

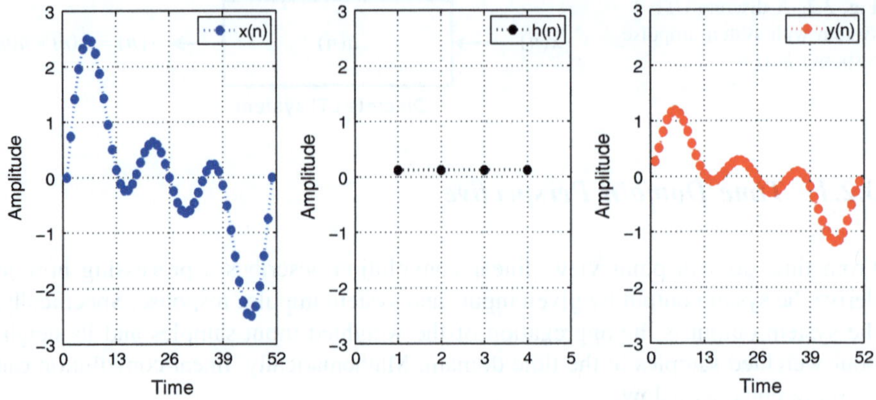

Fig. 3.3 An example of linear convolution from the time domain perspective

Figure 3.3 shows an example with some real numbers for **H** = [1/8, 1/8, 1/8, 1/8], which applies "moving average" effects to the input signal.

3.2.2 *Frequency Domain Perspective*

Sometimes, certain properties of the signal are "hidden" in its original time domain representation while easy-to-notice in the transformed domain, e.g., a periodic sine wave signal in the time domain will appear as a very simple single-tone signal in the frequency domain. Therefore, the same signal may look like contain "more" information in the transformed domain than its original time domain representation. And the direct benefit is that it will be relatively easier to analyze signals in the transformed domain for certain cases.

In digital signal processing, we use the so-called Z-transform to translate the real or complex discrete-time signal into complex frequency domain representation. Mathematically, Z-transform of a discrete-time signal $x(n)$ can be performed as Eq. 3.9.

$$X(z) = Z[x(n)] = \sum_{n=-\infty}^{\infty} x(n)z^{-n} \tag{3.9}$$

where $Z[\cdot]$ represents Z-transform function.

One of the most amazing properties of linear convolution is that, time domain linear convolution is equivalent to frequency domain multiplication operation. In other words, in order to perform the linear convolution in the time domain, we could carry out simple multiplication operation in the frequency domain. Here, for the sake of simplicity, we directly give the following proof as Eq. 3.10. And we will explore the detailed advanced application of this property in Chap. 6.

$$
\begin{aligned}
Z[x(n) * h(n)] &= Z\left[\sum_{m=-\infty}^{\infty} x(m)h(n-m)\right] \\
&= \sum_{n=-\infty}^{\infty}\left[\sum_{m=-\infty}^{\infty} x(m)h(n-m)\right]\cdot z^{-n} \\
&= \left[\sum_{m=-\infty}^{\infty} x(m)\right]\cdot\left[\sum_{n=-\infty}^{\infty} h(n-m)z^{-n}\right] \\
&= \left[\sum_{m=-\infty}^{\infty} x(m)\right]\cdot\left[\sum_{l=-\infty}^{\infty} h(l)z^{-(l+m)}\right] \\
&= \left[\sum_{m=-\infty}^{\infty} x(m)z^{-m}\right]\cdot\left[\sum_{l=-\infty}^{\infty} h(l)z^{-l}\right] \\
&= X(z)H(z)
\end{aligned}
\tag{3.10}
$$

The relationship between time domain linear convolution and frequency domain multiplication is shown in the Fig. 3.4.

Figure 3.5 illustrates the linear convolution from frequency domain perspective, in this example, the convolution actually performs a filter function, i.e., the frequency components above 400 Hz will be suppressed, the level of the suppression is specified by the frequency domain system response *H*(z). Depending on the behavior of *H*(z), the filter can provide low-pass function (the frequency components below certain specification will maintain after convolution, while the frequency components above the specification will be suppressed), high-pass function, band-pass functions, or equalization functions for smoothing continuous frequency spectrum of the signals.

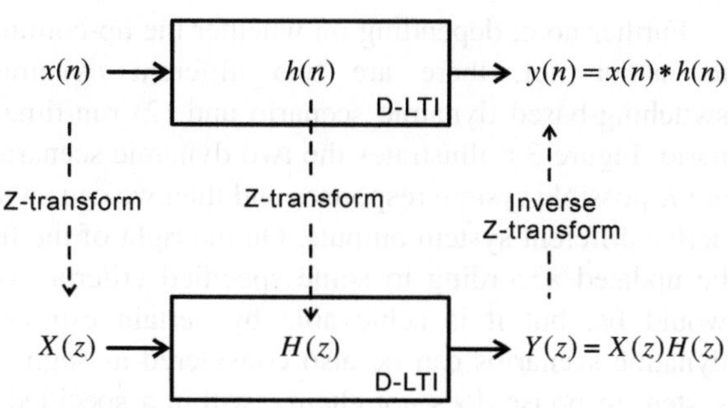

Fig. 3.4 Time domain linear convolution and its equivalent frequency multiplication

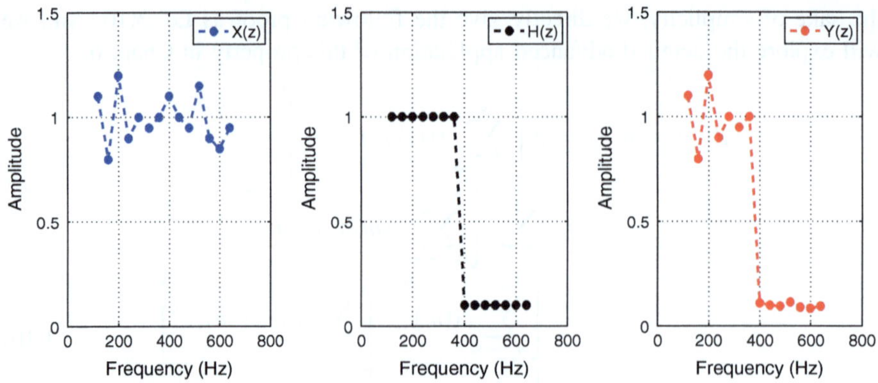

Fig. 3.5 An example of linear convolution from frequency domain perspective

3.2.3 Static and Dynamic Processing

If the system response $h(n)$ doesn't change with time, then the linear convolution between input and $h(n)$ can be considered as static processing, while if $h(n)$ will be changed/modified/updated/during the operation, then we consider this scenario as dynamic processing. For example, in the wireless transceivers, the linear convolution based low-pass FIR filter will be used at the baseband to remove the image components higher than a specified frequency. The system response of this FIR filter is characterized by $h(n)$. In the conventional transceivers, the baseband signal bandwidth is usually fixed, e.g., LTE with 25 resource blocks occupies 5 MHz bandwidth [1], thus the corresponding low-pass FIR can be fixed as well, i.e., $h(n)$ is static during the system operation. While in the software defined radio (SDR) transceivers, in order to achieve an efficient communication, the baseband processing will be carried out according to some dynamic requirement, for example, at time t_1, we are transmitting a 5 MHz LTE signal, and later at time t_2, we may want to transmit LTE signal with 100 resource blocks [1] to boost the data rate, which occupies 20 MHz bandwidth. For this dynamic scenario, the required response of low-pass FIR filter will be changed accordingly.

Furthermore, depending on whether the up-coming system responses are known to us or not, there are two different dynamic scenarios: (1) pre-known switching-based dynamic scenario and (2) run-time adaption-based dynamic scenario. Figure 3.6 illustrates the two dynamic scenarios. On the left, we know there are K possible system responses and then we can switch between those responses to derive different system outputs. On the right of the figure, the system response will be updated according to some specified criteria, we don't know exactly what it would be, but it is achievable by certain estimation approaches. Both of the dynamic scenarios can be also considered as segmental static processing, i.e., the system response does not change within a specified time period/time segment.

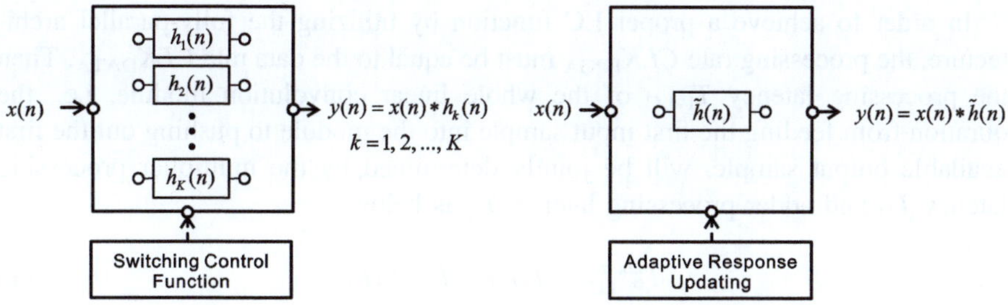

Fig. 3.6 Switching-based and adaption-based dynamic linear convolution

3.3 FPGA Implementation Architectures

In this section, we will have a look at how to build a linear convolution core in the FPGA platform. We intend to discuss the flexibility of FPGA implementation architectures, rather than providing only one optimized solution for achieving linear convolution on FPGAs. In other words, we focus on the worst case without the possibility of carrying out system level optimization. For instance, without loss of generality, we will focus on the asymmetric system response scenario, i.e., given an M-tap $h(n)$, it is not symmetrical with regards to the value of the taps, mathematically $h(n) \neq h(M - 1 - n)$. For $h(n) = h(M - 1 - n)$ scenario, since there are only half of the meaningful coefficients, architecturally the linear convolution for this scenario is quite similar to a linear convolution with $M/2$-tap (if M is an even number) or $(M + 1)/2$-tap (if M is an odd number) response.

3.3.1 *Fully-Parallel Architecture*

Firstly, let's have a look at the fully-parallel architecture to implement linear convolution. According to Eq. 3.6, the LC can be achieved as illustrated by Fig. 3.7 straightforwardly. In this architecture, the required number of multipliers is equal to the number of taps. In other words, we require M-multiplier and $(M - 1)$-adder to derive the outputs for an M-tap LC.

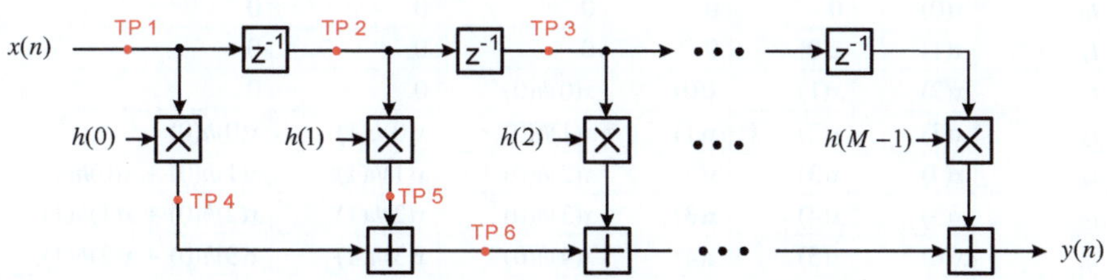

Fig. 3.7 Addition-chain-based fully-parallel LC implementation architecture

In order to achieve a proper LC function by utilizing the fully-parallel architecture, the processing rate CLK_{FPGA} must be equal to the data rate CLK_{DATA}. Then the processing latency $T_{P\text{-}LC}$ of the whole linear convolution module, i.e., the duration from feeding the first input sample into the module to pushing out the first available output sample, will be jointly determined by the multiplier processing latency T_M and adder processing latency T_A as below,

$$T_{\mathrm{P-LC}}^{chain} = T_M + (M - 1)T_A \tag{3.11}$$

T_M and T_A are usually characterized by the number of the processing clocks [2], i.e., how many cycles are required to produce the outputs after giving the inputs. Let's have a look in details. Table 3.1 illustrates the data processing flow (for the first 8 clock cycles, t_0 to t_7) in the fully parallel architecture at several test points (TP), and we assume $T_M = 2$ and $T_A = 1$, which means it will take 2 processing clocks for the multiplier and 1 processing clock for the adder to produce the corresponding outputs, respectively. In this example, TP1 shows the original input data samples. TP2 represents 1 unit-delayed version of TP1. Similarly, TP3 is 1 cycle delayed version TP2. TP4 and TP5 show the outputs of the first and the second multipliers, respectively. TP6 illustrates the output of the first adder, which performs the addition operation between TP4 and TP5.

Without any difficulty, we can recognize that the processing latency of the addition-chain-based fully-parallel architecture largely depends on how many adders are actually required. As the number of taps M increasing, the input to output latency of LC using the addition-chain-based fully-parallel architecture will increase proportionally according to the Eq. 3.11. An alternative solution, fully-parallel architecture with addition-tree approach can reduce the processing latency as below

$$T_{P-LC}^{tree} = T_M + [\log_2 M]_c T_A \tag{3.12}$$

where $[\cdot]_c$ represents ceiling function, i.e., the value will be round up to the nearest larger number. Function $[\log_2 M]_c$ denotes the number of the addition stages

Table 3.1 Data flow at different TPs for the addition-chain-based fully-parallel architecture

Time	TP1	TP2	TP3	TP4	TP5	TP6
t_0	$x(0)$	0	0	0	0	0
t_1	$x(1)$	$x(0)$	0	0	0	0
t_2	$x(2)$	$x(1)$	$x(0)$	$x(0)h(0)$	0	0
t_3	$x(3)$	$x(2)$	$x(1)$	$x(1)h(0)$	$x(0)h(1)$	$x(0)h(0)$
t_4	$x(4)$	$x(3)$	$x(2)$	$x(2)h(0)$	$x(1)h(1)$	$x(1)h(0) + x(0)h(1)$
t_5	$x(5)$	$x(4)$	$x(3)$	$x(3)h(0)$	$x(2)h(1)$	$x(2)h(0) + x(1)h(1)$
t_6	$x(6)$	$x(5)$	$x(4)$	$x(4)h(0)$	$x(3)h(1)$	$x(3)h(0) + x(2)h(1)$
t_7	$x(7)$	$x(6)$	$x(5)$	$x(5)h(0)$	$x(4)h(1)$	$x(4)h(0) + x(3)h(1)$

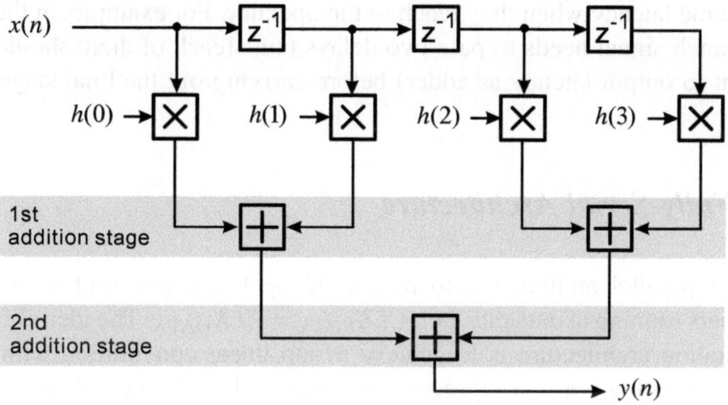

Fig. 3.8 Addition-tree-based fully-parallel LC implementation architecture (For the scenario: $\log_2 M$ is an integer)

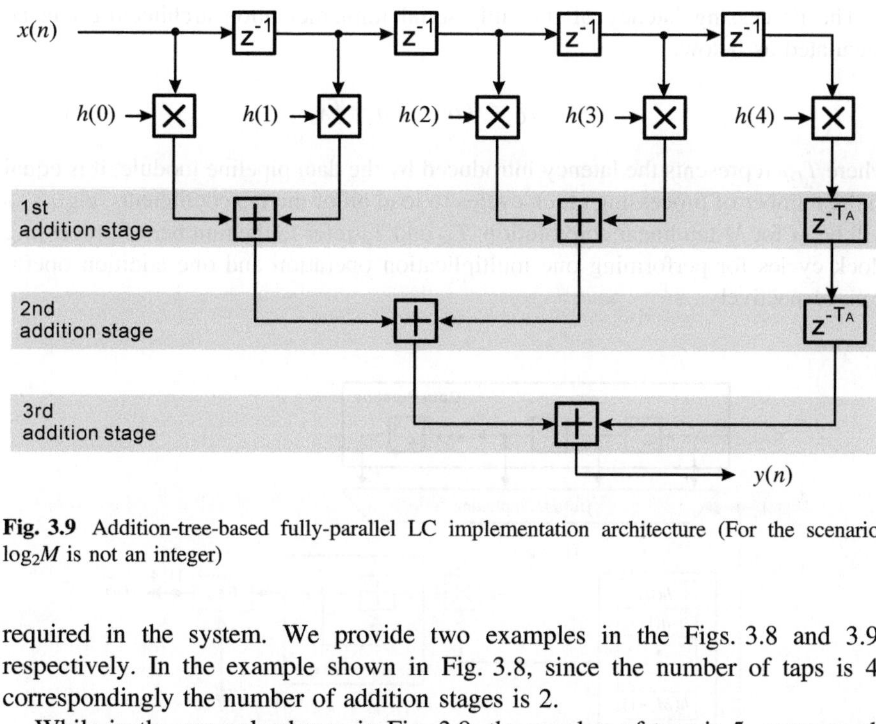

Fig. 3.9 Addition-tree-based fully-parallel LC implementation architecture (For the scenario: $\log_2 M$ is not an integer)

required in the system. We provide two examples in the Figs. 3.8 and 3.9, respectively. In the example shown in Fig. 3.8, since the number of taps is 4, correspondingly the number of addition stages is 2.

While in the example shown in Fig. 3.9, the number of taps is 5, correspondingly the number of addition stages is 3. In order to make it working properly, the adder latency needs to be considered to transfer the proper signal to the next stages. The general rule is the signals that will be processed by the same stage should have the same input to output latency. In other words, if we need to do basic arithmetic of two signals correctly at the intermediate processing stages, those two signals should have the same latency when they reach to the operator. For example, in the Fig. 3.9, the last branch signal needs to pass two delays units (each of them should have the same input to output latency as adder) before carrying out the final stage addition.

3.3.2 Fully-Serial Architecture

Using fully-parallel architecture to realize M-tap linear convolution will require M multipliers running at data rate, i.e., $CLK_{\text{FPGA}} = CLK_{\text{DATA}}$. The idea of fully-serial implementation architecture is to achieve M-tap linear convolution with only one multiplier running at a processing rate of M-time data rate, i.e., $CLK_{\text{FPGA}} = M \times CLK_{\text{DATA}}$. Figure 3.10 illustrates the fully-serial implementation architecture, besides one multiplier, there are six other modules including data pipeline module, data multiplexing (DM) module, coefficient random access memory (RAM) module, accumulator module, cycle detect module and one capture register.

The processing latency of the fully-serial implementation architecture can be calculated as below,

$$T_{S-LC} = T_{DP} + T_M + T_A \tag{3.13}$$

where T_{DP} represents the latency introduced by the data pipeline module, it is equal to the number of processing clock cycles to load all of the tap coefficients, e.g., T_{DP} will be M for M-tap linear convolution. T_M and T_A refer to the number of processing clock cycles for performing one multiplication operation and one addition operation, respectively.

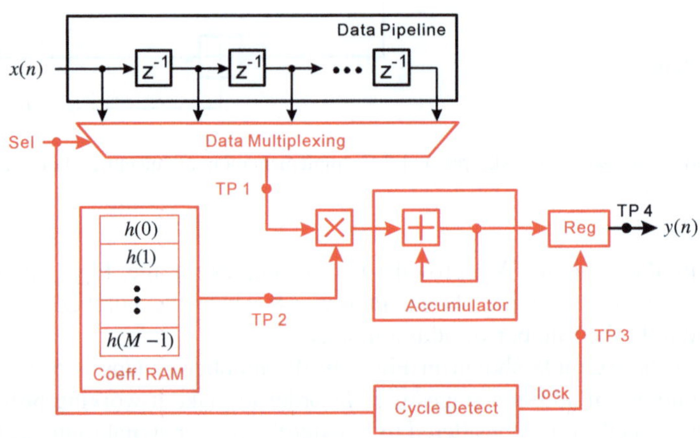

Fig. 3.10 Data multiplexing—based fully-serial LC implementation architecture

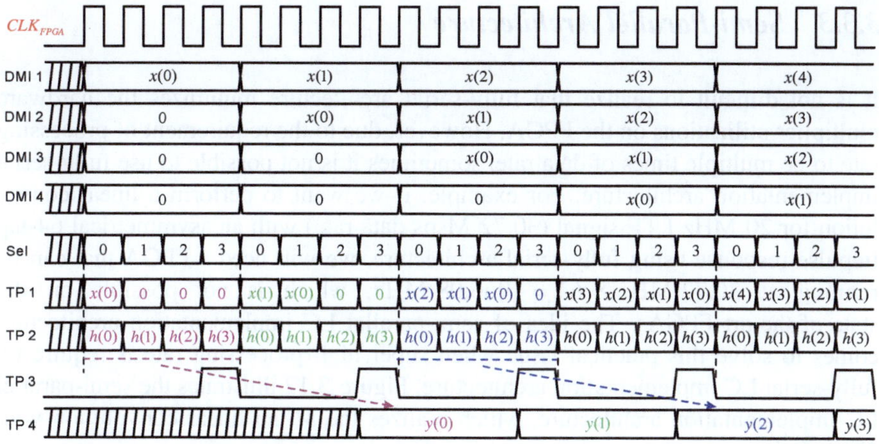

Fig. 3.11 Timing diagram of data multiplexing—based fully-serial LC implementation architecture (using 4-tap coefficients as an example)

A timing diagram shown in Fig. 3.11 is provided to illustrate how the LC is carried out by fully-serial architecture. For the sake of simplicity, we assume $M = 4$, i.e., a linear convolution with 4 taps (coefficients). DMI1 to DMI4 refer to the inputs to the DM module and they are simply delayed version of the original input $x(n)$. Signal Sel is a control signal (for DM module) periodically countering at processing rate, e.g., in this example, it is a 2-bit counter running at four times of data rate. This Sel signal guides the DM module to select the input signals (DM1 to DM4) to the DM module output. Then the output of DM module is illustrated by TP1. Simultaneously, the coefficient RAM module is pushing the coefficients periodically and consistently at the processing rate, illustrated by TP2. The accumulator aggregates the multiplied signal in a cycle of M, the value of which is 4 in the example. We assume the hardware multiplier and adder processing latency $T_M = 2$ and $T_A = 1$, respectively. Cycle detect module is to generate a lock or enable signal to capture the accumulated signal every four cycles. At last, the register will latch the convoluted results to the output.

Let's have a look at what happened to the data flow mathematically, illustrated by Eq. 3.14, which expands linear convolution definition of Eq. 3.6.

$$\begin{array}{rclclclcl} y(0) & = & x(0)\cdot h(0) & + & 0\cdot h(1) & + & 0\cdot h(2) & + & 0\cdot h(3) \\ y(1) & = & x(1)\cdot h(0) & + & x(0)\cdot h(1) & + & 0\cdot h(2) & + & 0\cdot h(3) \\ y(2) & = & x(2)\cdot h(0) & + & x(1)\cdot h(1) & + & x(0)\cdot h(2) & + & 0\cdot h(3) \\ y(3) & = & x(3)\cdot h(0) & + & x(2)\cdot h(1) & + & x(1)\cdot h(2) & + & x(0)\cdot h(3) \\ y(4) & = & x(4)\cdot h(0) & + & x(3)\cdot h(1) & + & x(2)\cdot h(2) & + & x(1)\cdot h(3) \\ \vdots & = & \vdots & + & \vdots & + & \vdots & + & \vdots \end{array} \tag{3.14}$$

3.3.3 Semi-Parallel Architecture

It is not difficult to realize that fully-serial architecture minimizes the hardware multiplier utilizations on the FPGA. However, due to the requirement of processing rate to be multiple times of data rate, sometimes it is not possible to use fully-serial implementation architecture. For example, if we want to perform a linear convolution for 20 MHz LTE signal (30.72 Msps data rate) with an asymmetrical 64-tap impulse response using fully-serial architecture, we will need a FPGA that can be running at 30.72 MHz × 64 = 1966.08 MHz, which is not possible on the state-of-the-art FPGAs. The idea of semi-parallel LC implementation architecture comes to solve this practical issue when super high processing rate is require by fully-serial LC implementation architecture. Figure 3.12 illustrates the semi-parallel LC implementation architecture, which realizes the *M*-tap linear convolution with *K* parallel segments.

In the semi-parallel architecture, the processing rate and data rate relationship is characterized by Eq. 3.15, which tells us that the requirement for the processing rate will be reduced proportionally to the number of parallel processing stages *K*. Instead of the *M* multipliers, *K* multipliers will be utilized, each of which is running at the processing rate of $CLK_{\mathrm{DATA}} \times M/K$.

$$\frac{CLK_{\mathrm{FPGA}}}{CLK_{\mathrm{DATA}}} = \frac{M}{K} \tag{3.15}$$

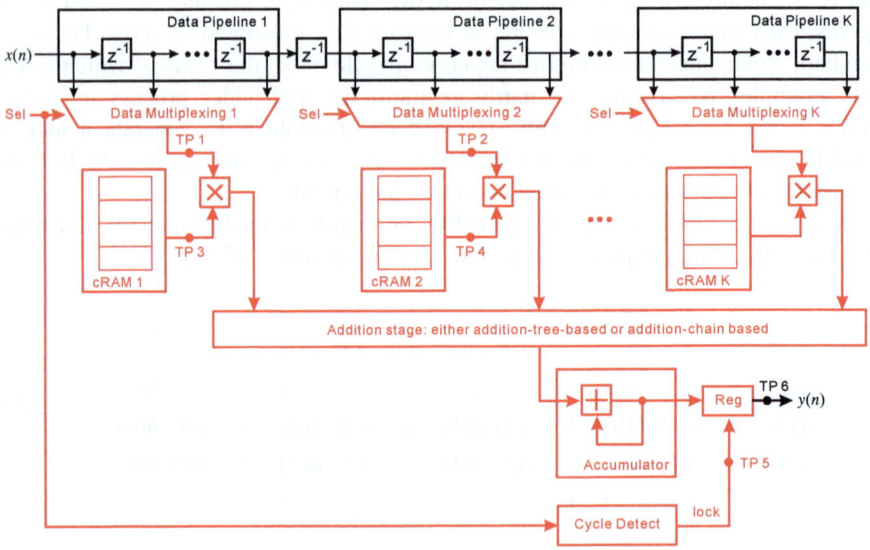

Fig. 3.12 Data-multiplexing- based semi-parallel LC implementation architecture

According to the architecture, the processing latency of the semi-parallel implementation architecture can be calculated as below,

$$T_{SP-LC} = T_{DP} + T_M + \lceil \log_2 K \rceil_c T_A + T_A \tag{3.16}$$

where T_{DP} represents the latency introduced by one data pipeline module, it is equal to the number of processing clock cycles to load $1/K$ of the tap coefficients, T_M and T_A refer to the number of processing clock cycles for performing one multiplication operation and one addition operation, respectively. Here we assume addition-tree architecture is used at the addition stage.

We provide the timing diagram as shown in Fig. 3.13 to illustrate how the semi-parallel architecture works, using a similar 4-tap linear convolution (M = 4, and K = 2) as an example. DMI 1-1 and DMI 1-2 refer to the two inputs to data multiplexing module 1, and DMI 2-1 and DMI 2-2 refer to the two inputs to data multiplexing module 2. We assume the hardware multiplier and adder processing latency T_M = 2 and T_A = 1, respectively.

The essential part of semi-parallel implementation architecture is how to properly determine the number of parallel stages K. First, we need to know the highest reliable processing rate of the multipliers on a specific FPGA device. For example, in the Xilinx data sheet for 7 series FPGA [3], the maximum processing rate of the multipliers, i.e., DSP48, is 650 MHz for -2 speed grade. Then according to the Eq. 3.15, given the value of the M, CLK_{DATA}, the maximum CLK_{FPGA}, we can derive the minimum number of the parallel stages K. In practices, if your projects / tasks are not in a tight resource constrained situation, we recommend "back-off" a

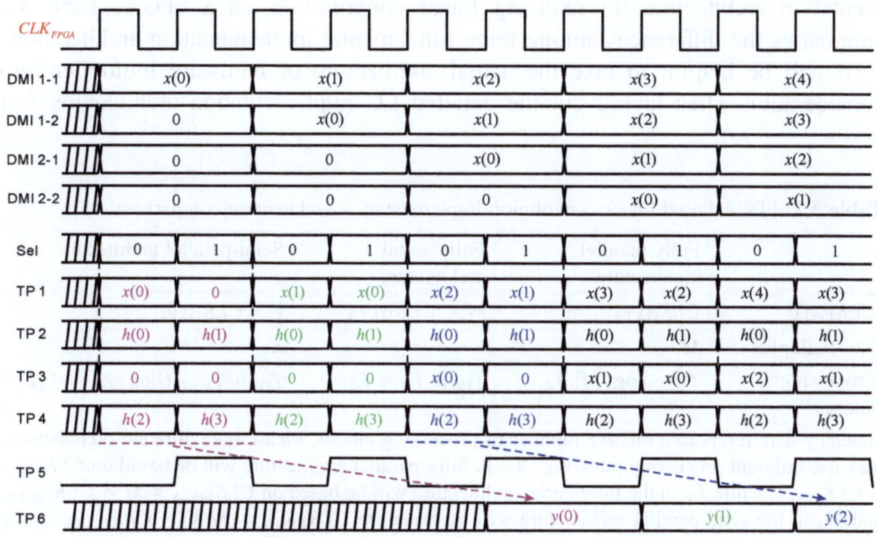

Fig. 3.13 Timing diagram of data multiplexing—based semi-parallel LC implementation architecture (using 4-tap coefficients as an example, M = 4, K = 2)

little bit from the highest clock rate, i.e., properly increase the K rather than using the minimum value (sometimes, the minimum integer value configuration already produce a "back-off" effect). Let's re-call the issue introduced early in this section, i.e., if we want to perform a linear convolution for 20 MHz LTE signal (30.72 Msps data rate) with an asymmetrical 64-tap impulse response, how many parallel stages do we need by using semi-parallel implementation architecture. The minimum number of parallel stages $K_{\min}$ can be derived by

$$K_{\min} = \left\lceil \frac{CLK_{\mathrm{DATA}}}{CLK_{\mathrm{FPGA}}^{\max}} \cdot M \right\rceil_c = \left\lceil \frac{30.72}{650} \times 64 \right\rceil_c = \lceil 3.02 \rceil_c = 4 \tag{3.17}$$

If we choose $K = 4$, then FPGA processing rate CLK_{FPGA} will be

$$CLK_{\mathrm{FPGA}} = \frac{M}{K} \cdot CLK_{\mathrm{DATA}} = \frac{64}{4} \times 30.72 = 491.52\ \mathrm{MHz}. \tag{3.18}$$

In this example, due to the ceiling function required derive an integer K, the minimum integer value K will lead to a processing rate that already backs-off from the highest available processing rate. By using semi-parallel architecture, we will use only 4 multipliers to perform 64-tap linear convolution operation.

3.3.4 Architecture Comparison

Depending on the system level requirement, we should pick up suitable implementation architecture for realizing linear convolutions on a FPGA. Table 3.2 summaries the differences among three fundamental implementation architectures.

It will be helpful to take the actual architecture of hardware multiplier into consideration when laying out the detailed LC implementation architecture. For

Table 3.2 FPGA-based linear convolution implementation architecture comparison

	Fully-parallel architecture	Fully-serial architecture	Semi-parallel architecture
CLK_{FPGA}	CLK_{DATA}	$M \times CLK_{\mathrm{DATA}}$	$\frac{M}{K} \times CLK_{\mathrm{DATA}}$
No. multipliers	M	1	K
Processing latency[1]	$T_M + \lceil \log_2 M \rceil_c T_A$	$T_{DP} + T_M + T_A$	$T_{DP} + T_M + (\lceil \log_2 K \rceil_c + 1) T_A$

[1]Latency here refers to a relative number of processing clocks. Please note different architectures may use different processing rates, e.g., T_M in fully-parallel architecture will be based on $CLK_{\mathrm{FPGA}} = CLK_{\mathrm{DATA}}$, while T_M in the fully-serial architecture will be based on $CLK_{\mathrm{FPGA}} = M \times CLK_{\mathrm{DATA}}$, and T_M in the semi-parallel architecture will be based on $CLK_{\mathrm{FPGA}} = (M/K) \times CLK_{\mathrm{DATA}}$ so the absolute latency will be different even given the same number of T_M

example, in Xilinx devices, the hardware multiplier DSP48 has already absorbed adder and pre-adder into the module [2]. Fully utilizing those existing resources will provide a "best suitable" solution for your applications.

3.4 Applications in Wireless Communications

In wireless communications, linear convolution has been widely utilized as digital filters, which can be categorized into two groups: digital filters for static processing and digital filters for dynamic processing.

In the digital front ends (DFE) of modern transceivers, usually we will put some multi-rate processing units in place for facilitating the whole transmission procedures. Those multi-rate processing modules convert the original data rate to a "proper" sampling rate at transmitter and vice versa at the receiver. The digital up convertor (DUC) in the transmitter and digital down convertor (DDC) in the receiver are multi-rate processing approaches, and usually the sampling rate will not change once it has been selected, which refers to the static processing. More precisely, the digital filter coefficients in those cases will not be updated in general.

Furthermore, in DFE, usually there is pre-equalization (pe-EQ) filter at transmitter and a post-equalization (post-EQ) filter at receiver for improving the non-flat frequency response introduced by the physical RF-chains (the nonlinearities introduced by RF power amplifiers will be discussed separately in Chap. 4). Those equalization filters are usually static in the conventional transceivers, since the transmission RF frequency is fixed, and the response of physical RF-chains are relative fixed as well, then the equalization coefficients can be pre-calculated and stored in some memories in the digital platform (e.g., FPGA) for flattening the frequency response consistently. However in the software-defined radio based transceivers, the carrier frequency can move from one frequency point to a different one, resulting in the requirement of re-calculation for the equalization coefficients to combat "new" non-flat frequency responses. This will be the dynamic processing which will update the equalization coefficients adaptively, and this approach is applicable for the wireless-channel related equalization as well.

Figure 3.14 shows a simplified function diagram of typical modern telecom wireless transceivers. Here we pick up the flexible integrated RF transceiver (which contains integrated DAC, ADC, analog mixers and some analog filters) as RF front end for example, and FPGA serves as digital signal processing platform.

3.4.1 Digital Up Conversion and Digital Down Conversion

Let's start the story with the linear convolution for static processing applications first. In wireless communications, the original baseband data rates are determined by the modulation schemes used in the system according to the particular wireless

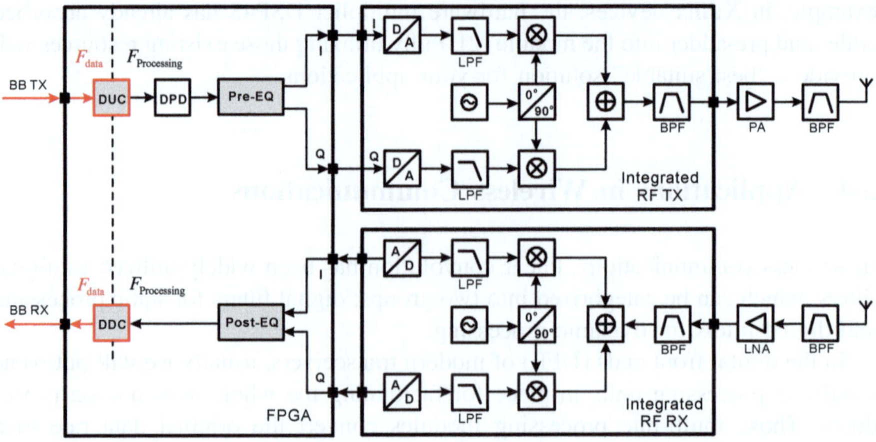

Fig. 3.14 Simplified diagram of modern telecom wireless transceivers (zero-IF TX and zero-IF RX architecture) using FPGA as digital platform and integrated RF TRX as analog solution

standards. For example, a 5 MHz LTE system will generate a complex baseband signal at data rate of 7.68 Msps [1]. Before doing the interaction between digital and analog domain by DAC at transmitter and ADC at receiver, usually at transmitter we will up-convert the baseband data rate signal to higher sampling rate data in the digital domain by the so-called up-sampling (interpolation) function (DUC), and at receiver we will capture the analog signal by ADC at the sampling rate that is more than required baseband data rate, and perform down-sampling (decimation) function (DDC). The main benefits of doing this multi-rate processing in the wireless transceivers are: (1) it facilities the data convertors providing high dynamic range signal in the analog domain; (2) it enables flexible transceiver architectures, such as complex intermediate frequency (IF) architecture, software-defined multi-standards single-chain wireless transceiver architecture.

At transmitter, the up-sampling function is mathematically achieved by Eq. 3.19, i.e., given an discrete sequence $x(n)$ of sampling rate F_x (assume $F_x = f_s$), up-sampling $x(n)$ by a positive integer factor N_U can be achieved by inserting $N_U - 1$ zeros between adjacent two samples of $x(n)$. The sampling rate of the new up-sampled signal $y(m)$ will be $F_y = F_x \times N_U$.

$$y(m) = \begin{cases} x\left(\frac{m}{N_U}\right) & m = 0,\ \pm N_U,\ \pm 2N_U,\ \ldots \\ 0 & \mathit{others} \end{cases} \tag{3.19}$$

A simplified system diagram of up-sampling together with anti-imaging filter (low-pass filter) is shown in Fig. 3.15. And Fig. 3.16 illustrates why we need the anti-imaging filter from frequency domain point view, here we use $N_U = 4$ as a simple example. The signal Nyquist bandwidths are shown in red thicker line, and $X(f)$, $Y(f)$, $H(f)$, $V(f)$ represent the original signal, up-sampled signal, anti-imaging filter response and output signal, respectively.

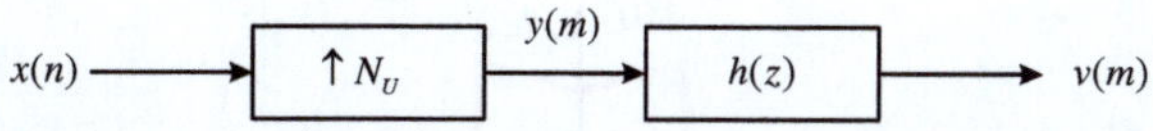

Fig. 3.15 A simplified function diagram of an up-sampling with anti-imaging filter

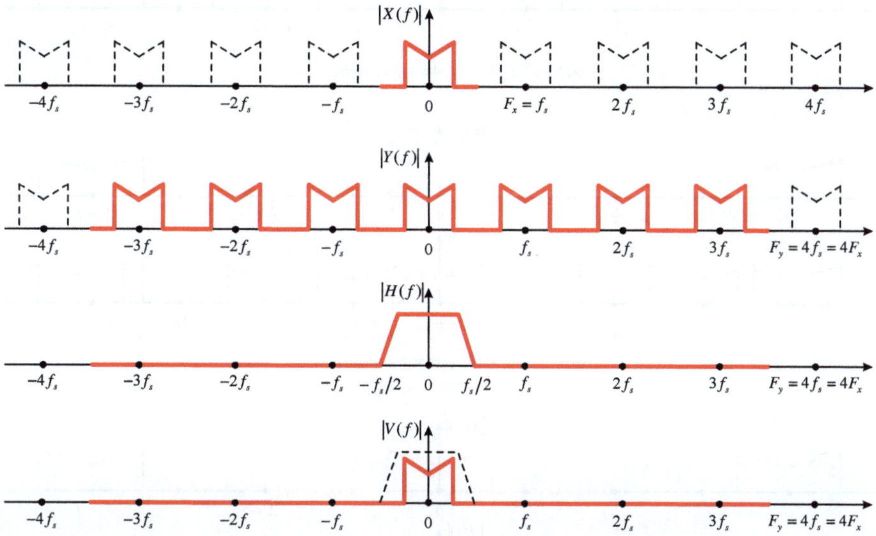

Fig. 3.16 Up-sampling and anti-imaging filter function in the frequency domain

At receiver, the down-sampling function is mathematically achieved by Eq. 3.20, i.e., given an discrete sequence $x(n)$ of sampling rate F_x ($F_x = f_s$), down-sampling $x(n)$ by a positive integer factor N_D can be achieved by extracting 1 sample every N_D original samples as a new sample of output. And the sampling rate of the new signal $y(m)$ will be $F_y = F_x \div N_D$.

$$y(m) = x(mN_D) \tag{3.20}$$

A simplified system diagram of down-sampling together with anti-aliasing filter is shown in Fig. 3.17.

And Fig. 3.18 illustrates why we need the anti-aliasing filters from frequency domain point view. Without anti-aliasing filter, the down-sampling processing may

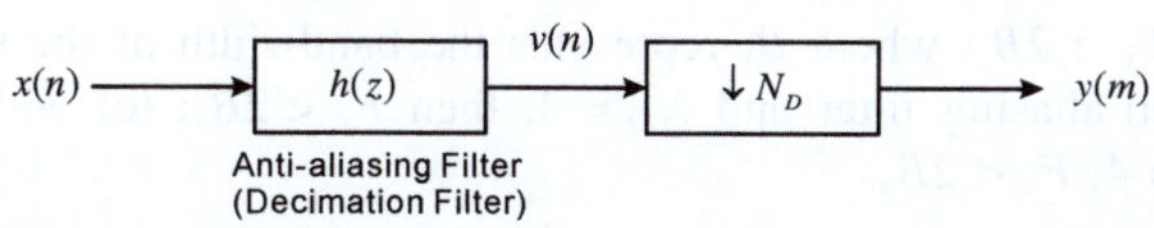

Fig. 3.17 A simplified function diagram of a down-sampling with anti-aliasing filter

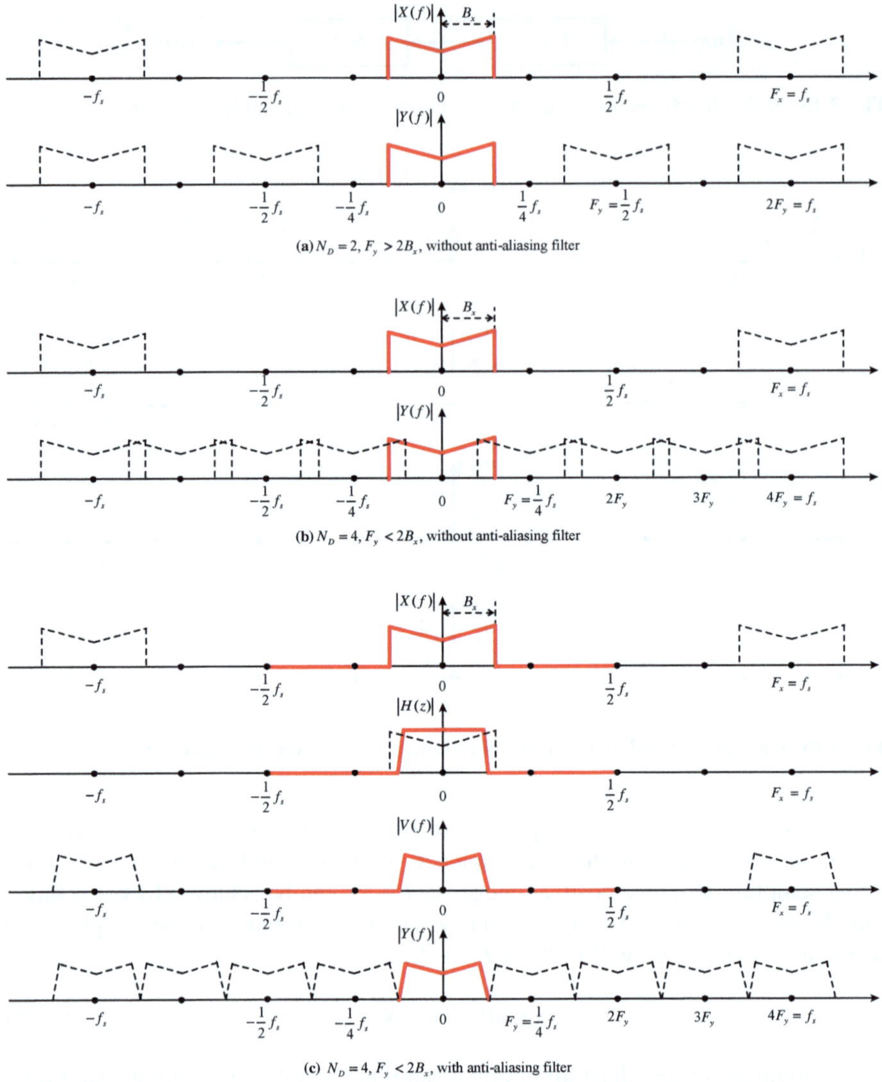

Fig. 3.18 Down-sampling and anti-aliasing filter function in the frequency domain

introduce aliasing effect in the frequency, i.e., the frequency spectrum in the multiple times sampling rate location will overlap with its neighbours.

In Fig. 3.18, we provide three scenarios: (a) without anti-aliasing filter and $N_D = 2$, then $F_y \geq 2B_x$, where B_x represents the bandwidth of the sequence $x(n)$; (b) without anti-aliasing filter and $N_D = 4$, then $F_y < 2B_x$; (c) with anti-aliasing filter and $N_D = 4$, $F_y < 2B_x$.

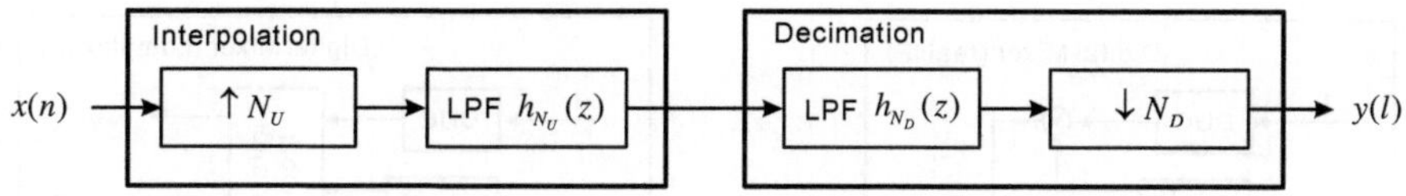

Fig. 3.19 A general function diagram of multi-rate sampling processing

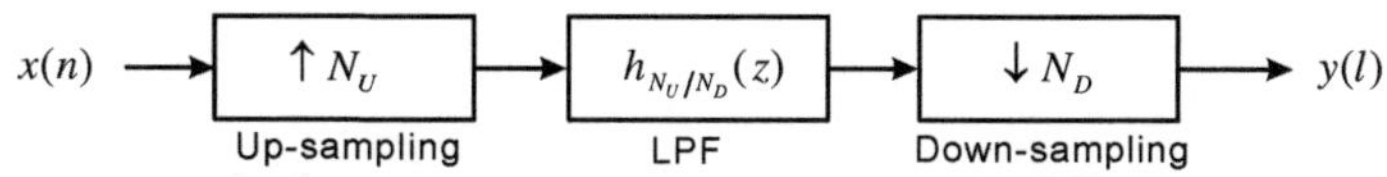

Fig. 3.20 Optimized general function diagram of multi-rate sampling processing (MRS)

Sometimes, fractional multi-rate processing may be required to achieve a desired processing rate, in this case, we just need to cascade the up-sampling and down-sampling units together. A general diagram of this is shown in Fig. 3.19 and since two linear convolutions are cascaded, we actually can use one linear convolution to represent the low-pass behavior as shown in Fig. 3.20.

Depending on the system level requirement, we can choose a suitable sampling rate for processing. For example, Table 3.3 provides some sampling rate settings used in the real systems for multi-rate processing in the 4G LTE system.

The usable baseband processing bandwidth is increased by using higher sampling rate, and some analog frequency mixing tasks can be flexibly carried out in the digital domain at baseband by the help of numerically controlled oscillator (NCO). Figure 3.21 illustrates the benefit of using high sampling rate regarding flexible configuration, where the digital mixer can produce real intermediate frequency (IF) outputs or complex IF outputs.

As modern DAC sampling rate is going higher and higher, using higher sampling rate at baseband is equivalent to having a larger baseband bandwidth, which can be used for transmitting multiple baseband signals simultaneously, in a carrier aggregation manner [4] as shown in Fig. 3.22, where multiple baseband signals of different data rates (7.68 Msps, 15.36 Msps, 30.72 Msps) are upconverted to 245.76 Msps and digitally mixed to particular IF. By utilizing high sampling rate processing, a single transmission chain will be adequate to handle the multiple independent signals of different standards.

Table 3.3 Multi-rate processing example in the modern wireless transceivers

Baseband signal LTE (MHz)	Data rate (Msps)	Exampled sampling rate configuration (MHz)											
		61.44		76.8		92.16		122.88		153.6		245.76	
		N_U	N_D	N_U	N_D	N_U	N_D	N_U	N_D	N_U	N_D	N_U	N_D
5	7.68	8	1	10	1	12	1	16	1	20	1	32	1
10	15.36	4	1	5	1	6	1	8	1	10	1	16	1
20	30.72	2	1	5	2	3	1	4	1	5	1	8	1

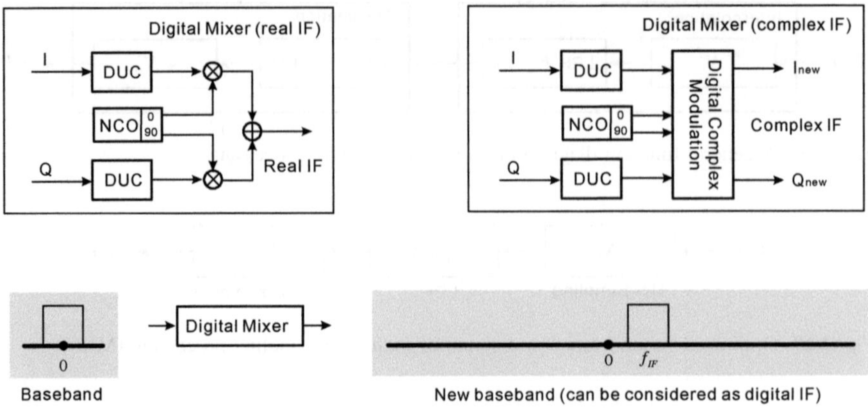

Fig. 3.21 DUC based digital mixer and its flexible frequency mixing function (it can be real IF architecture using one DAC at the next stage or complex IF architecture using two DACs at the next stage)

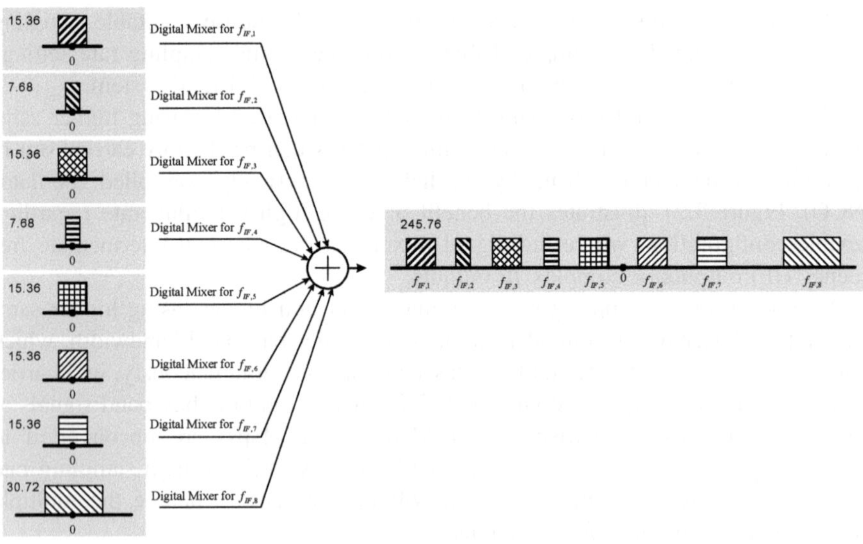

Fig. 3.22 Single transmission chain for multiple baseband signals using higher sampling rate

3.4.2 Frequency Pre-equalization and Post-equalization

The practical RF components are usually frequency sensitive devices and it is quite difficult to maintain the RF characteristics across a wide range in the frequency domain. In other words, the same device may produce different responses at different frequency bands, which is a quite common issue has to be deal with in the modern RF transceiver design, especially for the wideband RF transceivers. We can tune the RF components very hard trying to make it flat in terms of RF responses, or given a reasonable non-flat response situation, we can do something in the digital domain along the signal chain to combat this physical non-flat frequency response issue in an equalization manner. In the transmitter as shown in the Fig. 3.14, signal passes the equalization function first and then stimulates the RF stage, we name this type of equalization function as pre-equalization (Pre-EQ), while at the receiver side, the signal interacts with the RF stage first and then passes equalization function, we name this type of equalization function as post-equalization (Post-EQ).

The basic concept of equalization is not complicated, for example, a simplified diagram of pre-EQ enabled system is shown in Fig. 3.23, where h^{RF} and h^{EQ} (with M complex coefficients) represents the frequency response of RF stage and equalization filter, respectively. Signal $s(n)$ and $\tilde{s}(n)$ denote the original baseband complex signal and pre-equalized complex signal, respectively. With proper equalization, the cascade response from original digital signal to the RF radiating signal is flat in terms of spectrum characteristics.

The pre-equalization processing is mathematically described as the linear convolution between original signal and equalization filter response as below,

$$\tilde{s}(n) = s(n) * h^{EQ}(n) \tag{3.21}$$

Given the following complex representation, the actual pre-equalization processing is achieved by Eq. 3.23.

$$\begin{aligned} s(n) &= s_I(n) + js_Q(n) \\ \tilde{s}(n) &= \tilde{s}_I(n) + j\tilde{s}_Q(n) \\ h^{EQ} &= h_I^{EQ}(n) + jh_Q^{EQ}(n) \end{aligned} \tag{3.22}$$

$$\tilde{s}_I(n) + j\tilde{s}_Q(n) = [s_I(n) + js_Q(n)] * \left[h_I^{EQ}(n) + jh_Q^{EQ}(n)\right] \tag{3.23}$$

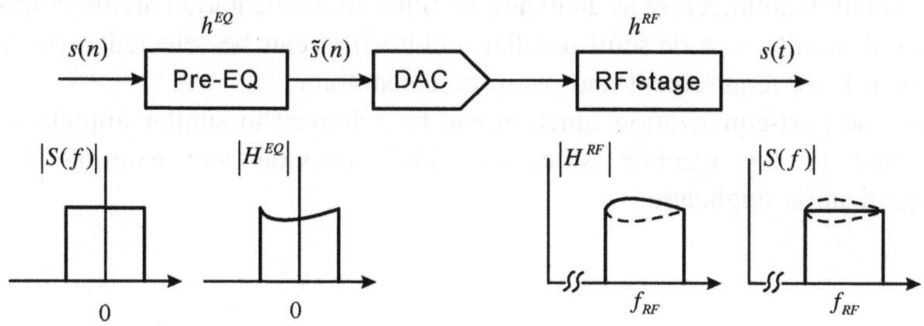

Fig. 3.23 The concept of equalization filter in the wireless transceiver

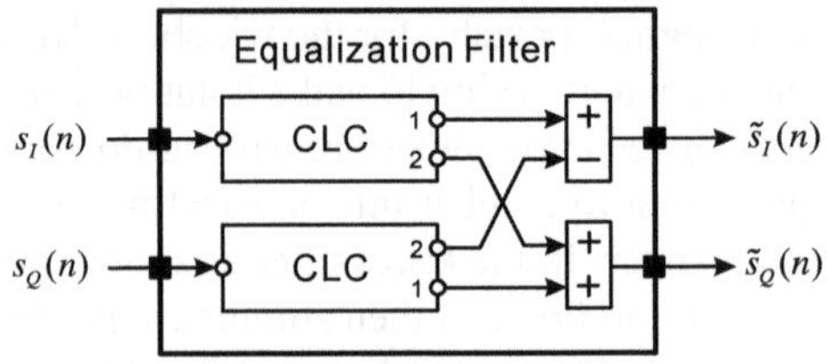

Fig. 3.24 Equalization filters using complex-coefficient linear convolution (CLC)

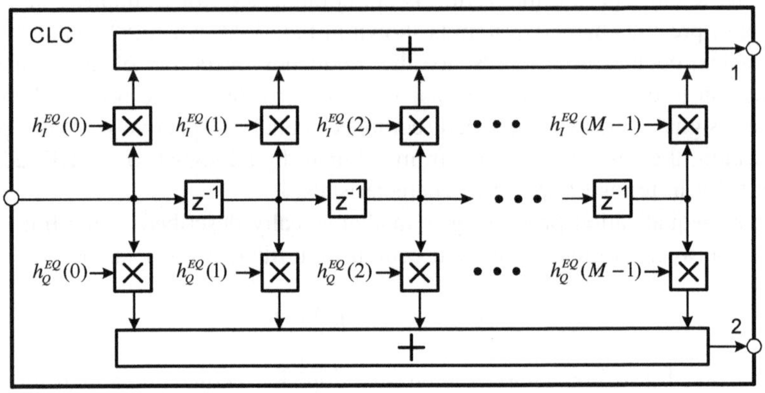

Fig. 3.25 Fully-parallel complex linear convolution implementation architecture

The pre-equalized $\tilde{s}_I(n)$ and $\tilde{s}_Q(n)$ can be represented as below, and the corresponding implementation architecture is shown in Fig. 3.24.

$$\begin{aligned}\tilde{s}_I(n) &= s_I(n) * h_I^{EQ}(n) - s_Q(n) * h_Q^{EQ}(n) \\ \tilde{s}_Q(n) &= s_I(n) * h_Q^{EQ}(n) + s_Q(n) * h_I^{EQ}(n)\end{aligned} \tag{3.24}$$

Figure 3.25 illustrates the CLC module implementation using fully-parallel implementation architecture as an example, other implementation architectures, like fully-serial architecture or semi-parallel architecture can be selected according to the system level requirement and resource constraints.

Since the post-equalization function can be achieved in similar implementation architecture at the receiver side, we don't provide any example for the post-equalization application.

3.4.3 Poly-phase Filter-based Interpolation for RF-DAC

As the technology moving forward, an RF-class digital to analog convertor (RF-DAC) has shown its powerful capability in the wireless applications by directly synthesizing the baseband digital signals to the RF (comparing to the traditional transmitter architecture using high-speed DAC with frequency mixers). The data rate of some RF-DAC has gone up to 6 GSPS, and the next generation would be 12 GSPS [5]. As mentioned before, interpolating the signal introduces images, and some anti-imaging filters are required to sufficiently suppress the images. The conventional interpolation function will perform up-sampling first and then carrier out the anti-imaging filtering function, which has to be running at higher sampling rate. This will be very difficult for RF-DAC devices, since the required sampling rate is at GSPS level, for example, 2.8 GSPS. Multi-phase linear convolution (poly-phase filtering) approach will significantly reduce the interpolation function design difficulties for RF-DAC applications. Let's have a look at an equivalent function regarding the interpolation as shown in Fig. 3.26.

Then alternative interpolation architecture using poly-phase decomposition can be derived as shown in Figs. 3.27, 3.28, 3.29 and 3.30.

The input-data-rate filters $h_p(z)$ ($p = 0, 1, \ldots, N_U - 1$) can be derived by splitting the high-speed filter coefficients into multiple subsets. For example, given $h(z^{N_U}) = [c_0, c_1, \ldots, c_n]$, then $h_p(z) = [c_p, c_{p+N_U}, \ldots]$. By using poly-phase decomposition and interpolation equivalent architecture, one output-data-rate filter (running at $N_u F_x$) has been replaced by multiple parallel input-data-rate filters (running at F_x). This using multiple parallel low-speed circuits to achieve serial high-speed function is quite aligned with the essential FPGA philosophy, and this

Fig. 3.26 Conventional interpolation (up-sampling first and filtering) and its equivalent interpolation function (filtering first and up-sampling)

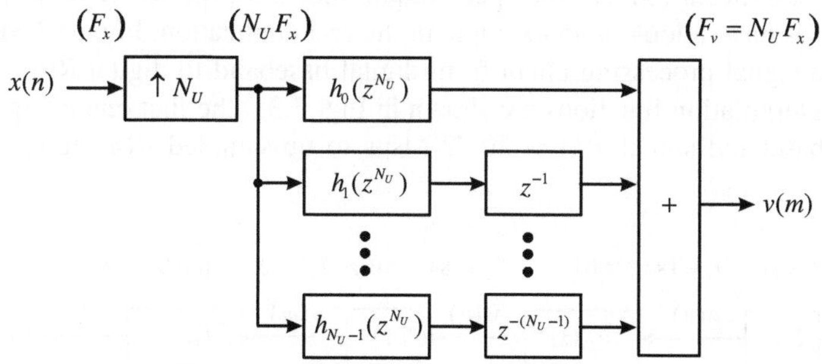

Fig. 3.27 Poly-phase decomposition of generalized interpolation function step 1

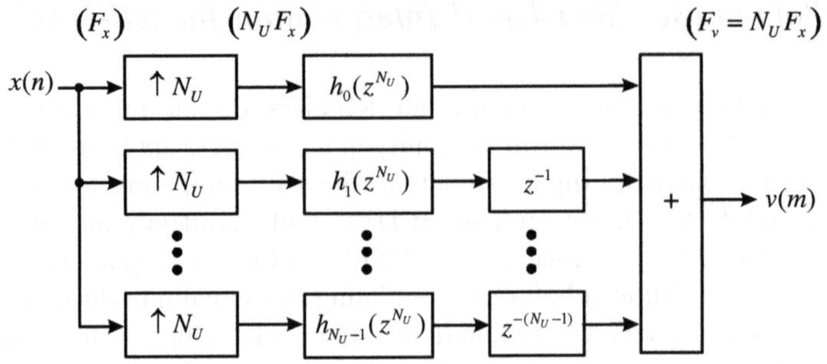

Fig. 3.28 Poly-phase decomposition of generalized interpolation function step 2

Fig. 3.29 Poly-phase decomposition of generalized interpolation function step 3

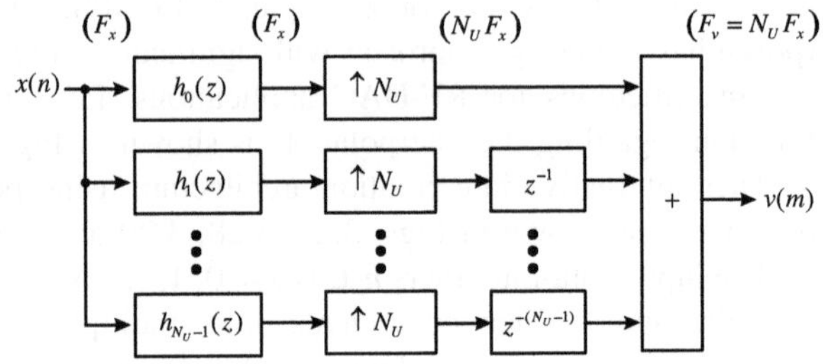

Fig. 3.30 Poly-phase decomposition of generalized interpolation function step 4

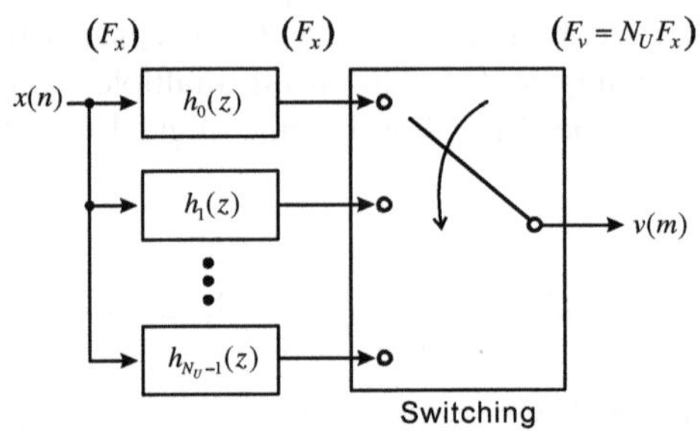

approach significant reduces the speed requirements for the RF-DAC interpolation filters. Let's have a look at an example in the real application. Figure 3.31 shows a simplified signal processing chain from digital baseband to digital RF.

Two interpolation functions are shown in Fig. 3.31, the first one interpolates the original baseband signal $x(k)$ at 30.72 Msps to up-sampled $w(n)$ at 184.32 Msps

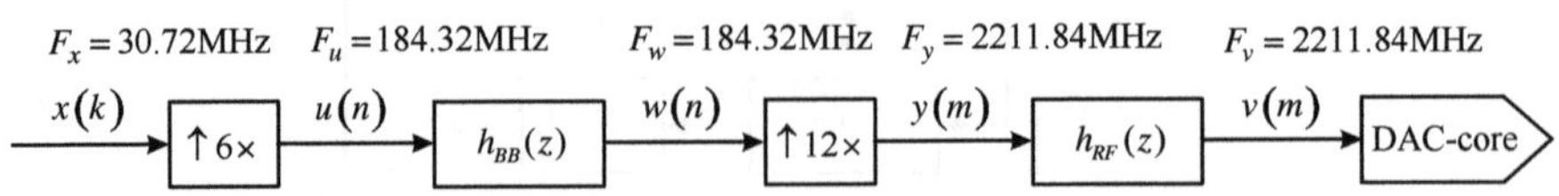

Fig. 3.31 Signal processing chain from digital baseband signal to digital RF signal

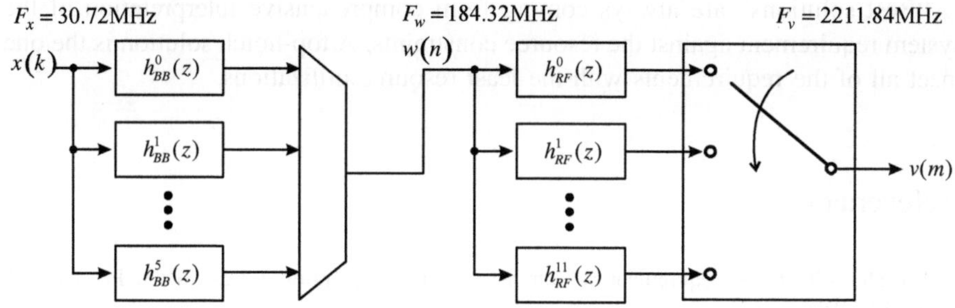

Fig. 3.32 Poly-phase based signal processing chain from digital baseband signal to digital RF signal

data rate. While the second interpolates further to RF sampling rate 2211.84 Msps. The LPF $h_{BB}(z)$ is running at 184.32 MHz, which can be achieved in the state-of-the-art FPGA, while using poly-phase decomposition can reduce the implementation requirements. The LPF $h_{RF}(z)$ is supposed to be running at 2211.84 MHz, which is beyond the processing speed of the current FPGAs, however we can achieve the same function using poly-phase decomposition based interpolation architecture running at much lower processing rate at the cost of parallel processing. For example, as shown in Fig. 3.32, we can achieve the same function the following poly-phase based architecture. The reduced sampling rate filters $h_{BB}^i(i = 0, 1, \ldots, 5)$ at 30.72 MHz and $h_{RF}^i(i = 0, 1, \ldots, 11)$ at 184.32 MHz can be simplified derived using sub coefficient sets from original data rate filters $h_{BB}(z)$ and $h_{RF}(z)$. Furthermore, after using parallel architecture, since each of the filters can be running at lower sampling rate, we can use semi-parallel or Fully-serial linear convolution architecture to implement those LPFs for making a proper trade-off between resources and processing speed.

3.5 Conclusions

In this chapter, we introduced the linear convolution concept from both time domain and frequency domain perspectives and its detailed FPGA implementation architectures including fully-parallel approach, fully-serial approach and semi-parallel approach. Comparison between different FPGA-based linear convolutions focused on the architecture complexity, processing rate to data rate ratio and processing latency. The applications of linear convolutions were illustrated by several fundamental modules on the modern wireless transceivers including digital up convertor (DUC), digital down convertor (DDC), equalization filters. The FPGA vendors, like Xilinx, usually provide some ready-to-use IP-cores of those functions and you will gain more information by reading some references [6].

"Best solutions" are always coming from comprehensive interpretation of the system requirement against the resource constraints. A top-notch solution is the one meet all of the requirements with the least resource utilizations.

References

1. E-UTRA (2011) user equipment radio transmission and reception, release 10, 3GPP standard, TS 136.101
2. Xilinx (2015) DSP48 Marco v3.0. LogiCORE IP Product Guide, PG 148
3. Xilinx (2016) Virtex-7 data sheet: DC and AC Switching. Data Sheet, DS 183
4. E-UTRA (2016) carrier aggregation base station radio transmission and reception, release 10, 3GPP standard, TR 36.808
5. Analog Device Company website, http://www.analog.com
6. Xilinx (2015) DUC/DDC Compiler v3.0. LogiCORE IP Product Guide, PG 147

Chapter 4
FPGA-based Nonlinear Convolution

Abstract This chapter will focus on the topic of discrete nonlinear convolution and how to build it with the FPGA platforms. The contents include the nonlinear convolution basics, different FPGA implementation architectures and typical applications in wireless communications, such as behavioural modelling of RF devices and digital predistortion for RF power amplifiers.

4.1 Introduction

In a linear system, convolution describes a behaviour of a system, the output of which depends on the input and its linear delayed versions, while in a nonlinear system, convolution characterizes a behaviour of a system, the output of which depends on the input and its nonlinear delayed versions. Nonlinear convolution is a very powerful tool that has been utilized in different areas of the modern wireless systems, especially for modelling nonlinear physical devices and mitigating or compensating for the nonlinear imperfections of physical RF devices. One of the popular approaches utilized as a nonlinear convolution, i.e., Volterra series, was firstly introduced by the mathematician Vito Volterra in 1887 [1], and later in the middle of 19th century, Norbert Wiener started using Volterra series in nonlinear circuit analysis, particularly for describing the so-called “weakly nonlinear” behaviour with fading memory. “Weakly nonlinear” refers to a situation that the fundamental term, rather than nonlinear terms, will dominate the system behaviour. Now let’s have a closer look at the nonlinear convolution.

4.2 Nonlinear Convolution Basics

Equation 4.1 denotes a general representation of a nonlinear system, which can be considered as a kind of mapping function translating the original input $x(n)$ from one domain to another domain with output $y(n)$ in a nonlinear way,

L. Guan, *FPGA-based Digital Convolution for Wireless Applications*,
Springer Series in Wireless Technology, DOI 10.1007/978-3-319-52000-1_4

$$y(n) = f[x(n)] \tag{4.1}$$

By saying "nonlinear way", we actually mean the following relationship,

$$\begin{aligned} &given\ y_k(n) = f[x_k(n)] \\ &a_k y_k(n) \neq f[a_k x_k(n)] \quad where\ a_k \neq 0 \end{aligned} \tag{4.2}$$

Particularly, among different nonlinear functions, a polynomial-based representation has been widely utilized due to its straightforward format, such as the so-called power series. Equation 4.3 denotes a *P*th-order power series, which describes nonlinear behaviour of a system without memory. We name this type of systems as memoryless nonlinear system.

$$y(n) = f[x(n)] = \sum_{p=0}^{P} h_p x^p(n) \tag{4.3}$$

where $x(n)$ and $y(n)$ represent system input and output, respectively. And h_p represents the weighting factor for the *p*th-order term. For instance, Eq. 4.4 illustrates the 3rd-order power series. It means, the output of the system depends on the instantaneous input and its high-order nonlinear power terms, while the historical input samples do not contribute to the instantaneous system output samples.

$$y(n) = h_0 + h_1 x(n) + h_2 x^2(n) + h_3 x^3(n) \tag{4.4}$$

Combining linear convolution and nonlinear power series creates a nonlinear convolution behaviour, like a Volterra series, which can be utilized to describe a nonlinear time invariant system with fading memory. Mathematically, *P*th-order discrete time Volterra series with *M* memory length can be written as below

$$\begin{aligned} y(n) = &\sum_{m_1=0}^{M} h_1(m_1)x(n-m_1) \\ &+ \sum_{m_1=0}^{M}\sum_{m_2=0}^{M} h_2(m_1,m_2)x(n-m_1)x(n-m_2) \\ &+ \cdots \\ &\sum_{m_1=0}^{M}\sum_{m_2=0}^{M}\cdots\sum_{m_P=0}^{M} h_P(m_1,\ldots,m_P)x(n-m_1)x(n-m_2)\cdots x(n-m_P) \end{aligned} \tag{4.5}$$

We can re-write the Volterra series in a compact format as Eq. 4.6,

$$y(n) = \sum_{p=1}^{P} \sum_{m_1=0}^{M} \cdots \sum_{m_p=0}^{M} h_p(m_1, \ldots, m_p) \prod_{k=1}^{p} x(n - m_k) \tag{4.6}$$

where $h_p(m_1,\ldots, m_p)$ is named as the pth-order Volterra Kernel. In practical, the nonlinear order P and memory length M are usually chosen as finite numbers to reduce the computational complexity. However even though, the truncated Volterra series are still quite complicated because of its multiple-dimensional convolution architecture. For instance, given very simple parameters $P = 2$ and $M = 1$, the truncated Volterra series can be represented by Eq. 4.7 as below.

$$\begin{aligned} y(n) &= \sum_{p=1}^{2} \sum_{m_1=0}^{1} \sum_{m_2=0}^{1} h_p(m_1, m_2) \prod_{k=1}^{p} x(n - m_k) \\ &= h_1(0)x(n) + h_1(1)x(n-1) \\ &\quad + h_2(0,0)x^2(n) + [h_2(0,1) + h_2(1,0)]x(n)x(n-1) + h_2(1,1)x^2(n-1) \end{aligned} \tag{4.7}$$

4.2.1 Time Domain Perspective

From time domain point view, nonlinear convolution describes a processing how to derive the system output by given system response and inputs. Unlike in the memoryless nonlinear system, output of which does not depend on the historical samples, in the nonlinear system with memory effects (nonlinear convolution behaviour), historical samples (or called memory samples) contribute to the system output. Figure 4.1 illustrates a memoryless nonlinear system, the output of which depends on its instantaneous input sample (in the example, the output sample at time t_n is determined by the input sample at time t_n and system response).

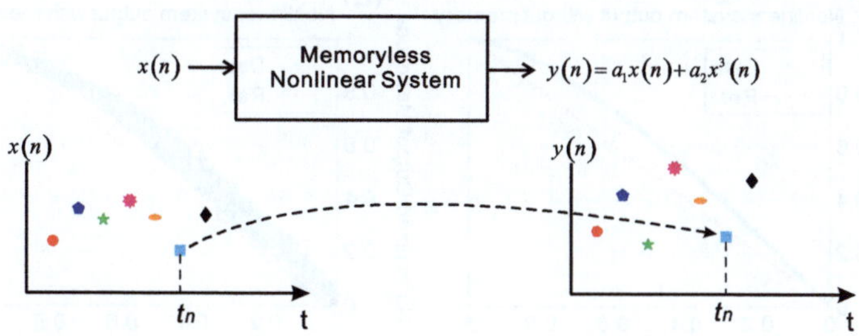

Fig. 4.1 Memoryless nonlinear system and its input-to-output mapping

Figure 4.2 shows an example of nonlinear system with memory effects, i.e., the output at time t_n not only depends on the instantaneous input at time t_n, but also depends on the historical inputs at time t_{n-1} and t_{n-2} (in this example, we assume memory length is 2).

Let's have a look at the difference between linear systems without and with memory effects, nonlinear system without and with memory effects. Figure 4.3 illustrates the four cases described above, where we passed the same 4G LTE 20 MHz signal (64 K complex samples) via four different systems and plotted the so-called AM-to-AM curves, i.e., normalized input magnitude to normalized output

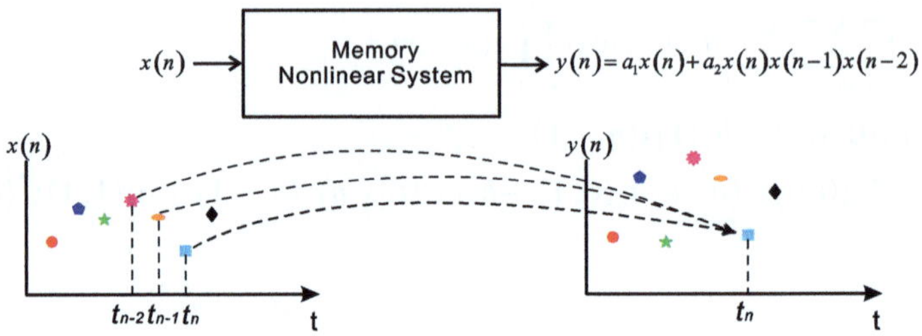

Fig. 4.2 Nonlinear system with memory effects and its input-to-output mapping

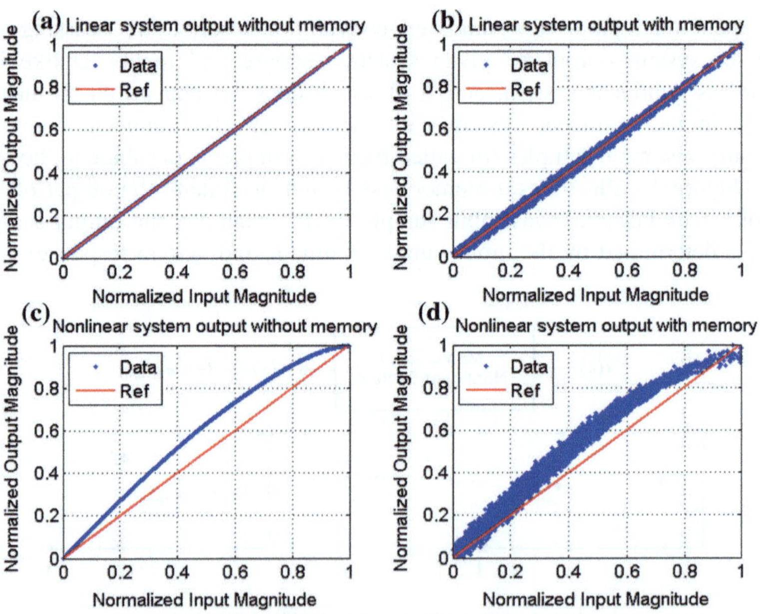

Fig. 4.3 AM-to-AM curves comparison among linear system without memory effects, linear system with memory effects, nonlinear system without memory effects and nonlinear system with memory effects, from time domain perspective

magnitude mapping, which describes the system transfer functions (input to output mapping) in the time domain. In each sub-figures, we provided a linear line as reference showing a linear input to output relationship. In Fig. 4.3a, linear system output without memory will behave like the same input signal after normalization. In Fig. 4.3b, after passing a linear system with memory effects (e.g., FIR filter), we observed a fussy thicker mapping in the AM-to-AM curve and this is because in the time domain, the input to output relationship is not one-to-one mapping anymore due to the memory effects. Figure 4.3c illustrates a memoryless nonlinear behaviour, which reflects the nonlinear one-to-one mapping in the time domain, while Fig. 4.3d shows the nonlinear behaviour with memory effects, which can be considered as the combination results of memoryless nonlinearities (Fig. 4.3c) and memory effects (Fig. 4.3b).

The physical active RF components, like power amplifiers, used in the wireless communications, usually suffer from the nonlinearities and memory effects [2, 3], which not only introduce the in-band distortion degrading the transmission signal itself but also cause out-of-band spectrum regrowth blocking the other signals at the adjacent frequency bands. Engineers and researchers have been trying to "linearize" the RF power amplifiers to improve the signal conditioning and transmission efficiency, and next let's have a look at the nonlinear convolution from frequency domain perspective.

4.2.2 Frequency Domain Perspective

From frequency domain perspective, nonlinear convolution describes a frequency dependent nonlinear behaviour, i.e., the signal at different frequencies will experience different levels of distortions. Let's re-visit the example used in the last section but this time we plotted the power spectral density (PSD) in Fig. 4.4 for the four cases showing the differences from frequency domain perspective.

A 20 MHz LTE signal was utilized to stimulate four different systems. In Fig. 4.4a, the signal passed via a linear system without memory effects, the output spectrum is the same as the one of its input signal, so for the sake of the simplicity we did not provide the spectrum of the original input signal, since it will be completely overlapped with the corresponding output spectrum. Figure 4.4b illustrates the case that we fed the signal to a linear system with memory effects, the output spectrum was shaped across the frequencies, i.e., different frequencies went through different-level distortion. As introduced before, the wideband signal passing via nonlinear system with or without memory effects will cause spectrum regrowth as shown in Fig. 4.4c, d, and with memory effects we observed frequency-dependent distortion behaviour in the frequency in Fig. 4.4d.

Figure 4.5 illustrates the PSD of the difference signal between output and input for four corresponding test scenarios. In linear system without memory effects (e.g., complex rotation), the normalized output minus normalized input will lead to certain in-band difference and the corresponding Error PSD is shown in Fig. 4.5a,

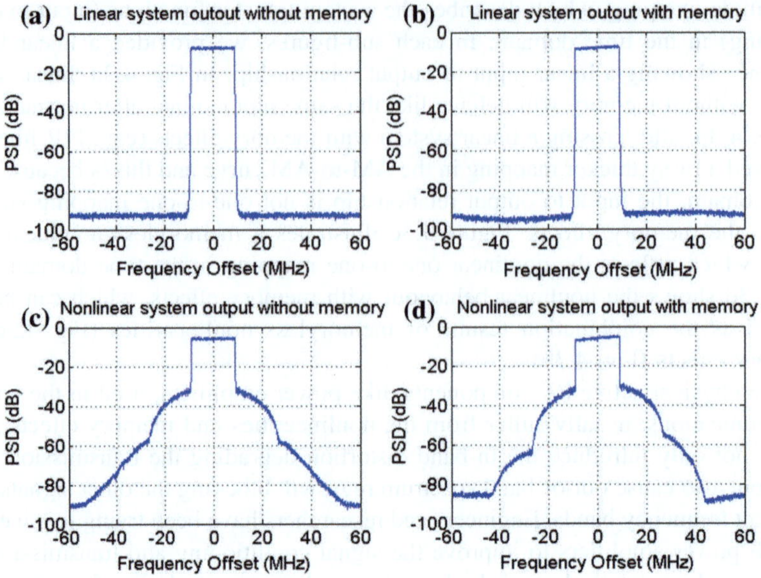

Fig. 4.4 Power spectrum density (PSD) plots of the output signals passing via linear system without memory effects, linear system with memory effects, nonlinear system without memory effects and nonlinear system with memory effects, from frequency domain perspective

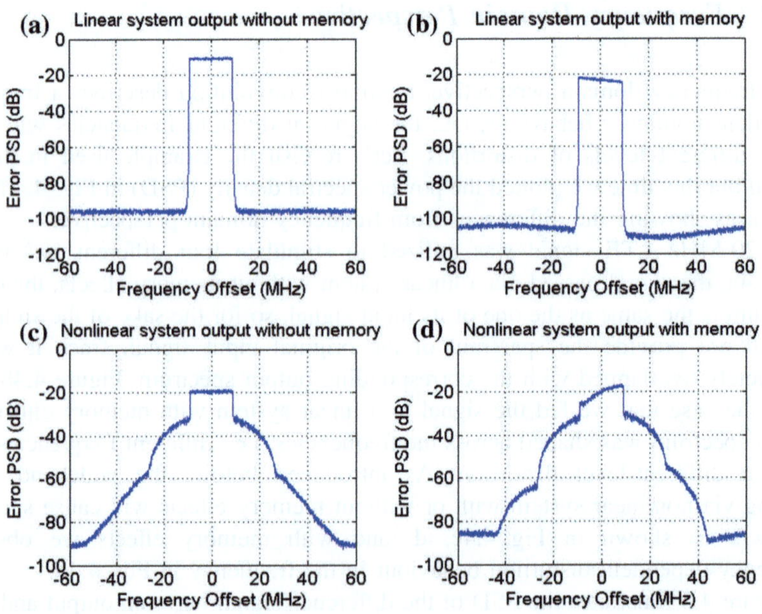

Fig. 4.5 Error PSD for the corresponding four scenarios: linear system without memory effects, linear system with memory effects, nonlinear system without memory effects and nonlinear system with memory effects, from frequency domain perspective

while in the linear system with memory, the frequency dependent behaviour is observed as shown in Fig. 4.5b, where within the frequency band, different frequency components went through different levels of distortions. Figure 4.5c, d illustrate the error PSD for the nonlinear behaviour without memory effects and with memory effects, respectively. And it is not difficult to notice that the in-band difference is quite flat in the frequency domain (non frequency dependent behaviour) in Fig. 4.5c while the in-band difference is non-flat in the frequency domain (frequency dependent behaviour, memory type behaviour) in the Fig. 4.5d.

The detailed frequency domain Volterra series analysis is out of the scope of this book, so we simply provide a definition of the corresponding Kernel in the frequency domain [4] as below.

$$H_p(j\omega_1, j\omega_2, \ldots, j\omega_p) = \int_{-\infty}^{\infty} \cdots \int_{-\infty}^{\infty} h_p(m_1, m_2, \ldots, m_p) e^{-j(\omega_1 m_1 + \cdots \omega_p m_p)} dm_1 \cdots dm_p \tag{4.8}$$

where H_p is named as frequency domain Volterra Kernel and it is obtained by a so-called p-dimensional Fourier transformation of the time domain pth order Volterra Kernel h_p. The detailed Volterra series mathematical representations in the frequency domain are available via the books, like [1, 4].

4.2.3 Static and Dynamic Processing

Similar to the linear convolution, the nonlinear convolution processing can be static or dynamic. If the nonlinear behaviour doesn't change with time, we consider this scenario as static nonlinear convolution, while if the nonlinear behaviour changes during normal operation, we name this scenario as dynamic processing. In the real application, the wireless systems are usually static within a period of time. Using nonlinear RF PA device as an example, when the active PA reaches to the thermal equilibrium status, the nonlinear behaviour the PA will be determined by four dominating factors: (a) device-dependent characteristics, (b) input signal bandwidth, (c) the average power of the input signal and (d) peak to average power ratio (PAPR) of the input signal. Then, if those four factors maintain the same during a period of the time, the nonlinear behaviour of the PA can be considered as static, while if we only change the input average power, the PA will behave a little bit different leading to a "new" nonlinear behaviour. Once the power of the input signal arrives at the new stable power level, we can consider the nonlinear behaviour of the PA is static again but in another form. Figure 4.6 illustrates two dynamic nonlinear convolution scenarios, one is based on switching between k different nonlinearities, and the other one is based on adaptive pth order nonlinearity.

The switching-based dynamic nonlinear convolution can be considered as time divided static processing, i.e., within different active periods of the time (assume the

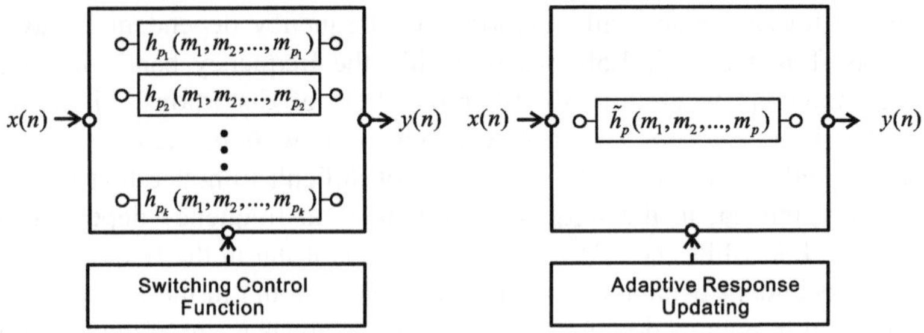

Fig. 4.6 Switching-based and adaption-based dynamic nonlinear convolution

actual switching time is close to zero), the generalized nonlinear kernel doesn't change, leading to the nonlinear behaviour maintain the same.

4.3 FPGA Implementation Architectures

Without loss of generality, a generalized Volterra series type nonlinear convolution can be constructed as shown in Fig. 4.7, which depicts a weighted nonlinear mapping from input data sequence $x(n)$ to output data sequence $y(n)$.

Memory mapping part produces M memory terms together with the original input term $x(n)$, and the generalized nonlinear mapping will select memory terms from this $(M + 1)$ collection like combination in mathematics, in which the order of selection does not matter and the same term can be selected multiple times. For example, given $M + 1$ memory terms (including the original term and M memory terms), there are $M + 1$ combinations of one that can be selected for $h_1(m_1)$ mapping. And there are $(M + 1) \times (M + 1)$ combinations of two (the same term can be selected twice) that can be drawn for $h_2(m_1, m_2)$. However, since $x(n)x(n - 1)$ and

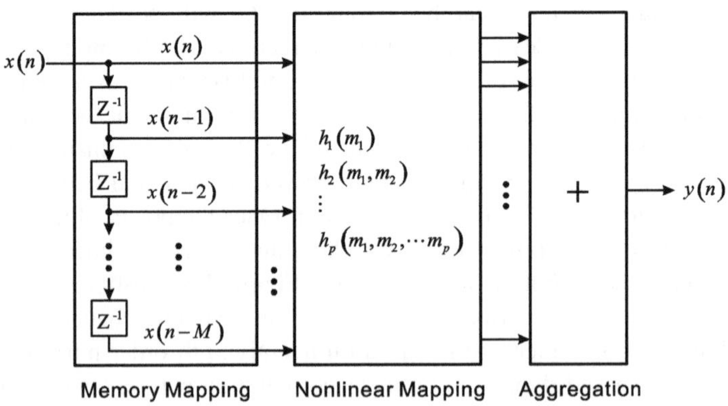

Fig. 4.7 Generalized nonlinear convolution with memory mapping and nonlinear mapping

$x(n-1)x(n)$ effectively are the same term, the corresponding weightings $h_2(0,1)$ and $h_2(1,0)$ can be combined. Let's define C_n^k as the number of k-combinations from a given set of n elements (the selecting order does not matter), and then the number of unique terms in the generalized nonlinear convolution (nonlinear order = P and memory length = M) can be derived as Table 4.1.

The number of unique pth order terms N_{GNC}^p in the generalized nonlinear convolution representation can be derived by the Eq. 4.9, where the coefficients $a_i(i = 1,\ldots,p)$ in this expansion are precisely the numbers on row $(p-1)$ of Yang Hui's triangle (or known as Pascal's triangle) as shown in Fig. 4.8.

$$N_{GNC}^p = a_1 C_{M+1}^p + a_2 C_{M+1}^{p-1} + a_3 C_{M+1}^{p-2} + \cdots a_p C_{M+1}^1 \tag{4.9}$$

In the wireless system, we usually utilize the complex representation, e.g., I and Q representation, to describe baseband signals. Adding complex baseband representation into the general Volterra series, we need to create another linear mapping first from complex baseband signal $x(n)$ to a constellation set $\{x(n), -x(n), x^*(n), -x^*(n)\}$ as shown in Fig. 4.9.

And if we absorb the sign of the signal to the corresponding weighting coefficients, we then reach to a mapping between $x(n)$ to a set of signal $\{x(n), x^*(n)\}$. When applying generalized nonlinear convolution to practical wireless applications, like baseband equivalent nonlinear modeling or baseband digital predistortion based nonlinear RF distortion compensation, we need to replace the $x(n)$ with $\{x(n), x^*(n)\}$ in the generalized nonlinear convolution representation. Full Volterra series,

Table 4.1 Number of unique terms in generalized nonlinear convolution representation

Order	Kernel	Number of unique terms N_{GNC}^p
1st	$h_1(m_1)$	C_{M+1}^1
2nd	$h_2(m_1, m_2)$	$C_{M+1}^2 + C_{M+1}^1$
3rd	$h_3(m_1, m_2, m_3)$	$C_{M+1}^3 + 2C_{M+1}^2 + C_{M+1}^1$
4th	$h_4(m_1, m_2, m_3, m_4)$	$C_{M+1}^4 + 3C_{M+1}^3 + 3C_{M+1}^2 + C_{M+1}^1$
5th	$h_5(m_1, m_2, m_3, m_4, m_5)$	$C_{M+1}^5 + 4C_{M+1}^4 + 6C_{M+1}^3 + 4C_{M+1}^2 + C_{M+1}^1$
…	…	…

Fig. 4.8 Yang Hui's triangle (Pascal's triangle) for the first 7 rows (row 0–6)

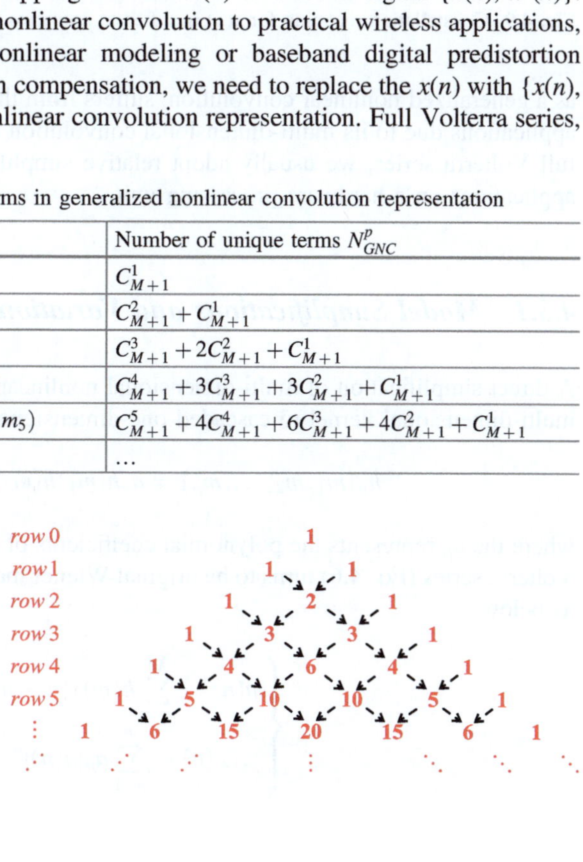

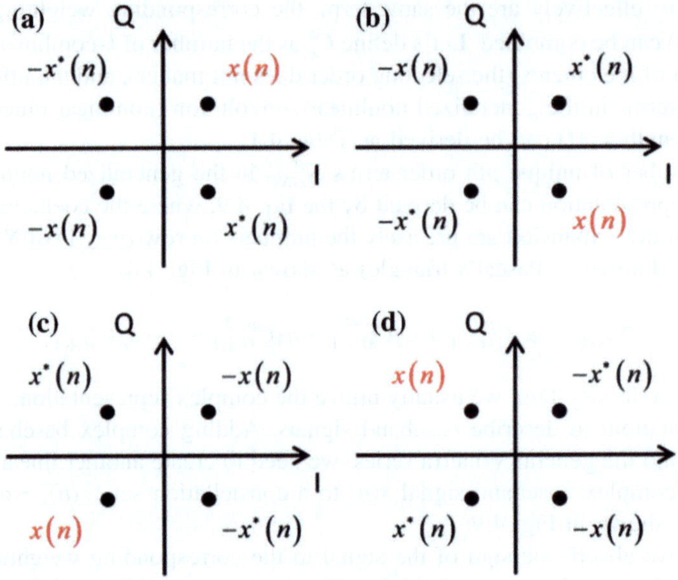

Fig. 4.9 Constellation mapping for $x(n)$ in different scenarios

as a generalized nonlinear convolution, suffers from the high complexity in the real applications due to its multi-dimensional convolution architecture. Instead of using full Volterra series, we usually adopt relative simplified architecture for practical applications on a hardware processing unit.

4.3.1 Model Simplifications and Variations

A direct simplification of multi-dimensional nonlinear convolution is to reduce the multi-dimensional kernel to cascaded one-dimensional kernel, e.g.,

$$h_p(m_1, m_2, \ldots, m_p) = a_p h(m_1) h(m_2) \cdots h(m_p) \tag{4.10}$$

where the a_p represents the polynomial coefficients of the nonlinearity, then the full Volterra series (Eq. 4.6) turns to be original Wiener model [5], which can be written as below,

$$\begin{cases} u(n) = \sum\limits_{m=0}^{M} h(m)x(n-m) \\ y_{OW}(n) = \sum\limits_{p=1}^{P} a_p [u(n)]^p \end{cases} \tag{4.11}$$

Original Wiener (OW) has very clear physical meaning due to its two-part model architecture, i.e., a simple $(M + 1)$-tap FIR filter followed by a static P-order nonlinear power series function. By expanding the FIR part, an Augmented Wiener (AW) model [6] was proposed to enhance the modelling accuracy and capability. The AW model can be written as,

$$\begin{cases} u(n) = \sum\limits_{m=0}^{M_1} h_1(m)x(n-m) + \sum\limits_{m=0}^{M_2} h_2(m)x(n-m)|x(n-m)| \\ y_{AW}(n) = \sum\limits_{p=1}^{P} a_p[u(n)]^p \end{cases} \tag{4.12}$$

Furthermore, if we generalize the FIR part to multiple FIR subsystems in parallel and using a generalized nonlinear function to replace nonlinear power series, we will reach to a parallel Wiener model [7] as illustrate by Eq. 4.13, which shares the same architecture of single-hidden-layer direct dynamic artificial neural network (DD-ANN) model [8, 9].

$$\begin{cases} u_k(n) = \sum\limits_{m=0}^{M} h_k(m)x(n-m) \\ y_{DD-ANN}(n) = \sum\limits_{k=1}^{K} a_k f[u_k(n)] \end{cases} \tag{4.13}$$

The ANN model is well known for its universal approximation capability (learning capability), especially its accurate global approximation of a strongly nonlinear system. Like convolutional neural network (CNN)—based deep learning approach [10, 11] has drawn a lot attention as a base machine learning algorithm for future ultimate artificial intelligence. Besides the DD-ANN model, sometimes we use recursive dynamic ANN (RD-ANN) [12] model that takes the delayed output into account, which means we use the current output of the system as part of inputs for deriving the next output for the future. An example of RD-ANN with a single-hidden-layer can be written as Eq. 4.14.

$$\begin{cases} u_k(n) = \sum\limits_{m=0}^{M} h_k(m)x(n-m) + \sum\limits_{m=1}^{M} g_k(m)y(n-m) \\ y_{RD-ANN}(n) = \sum\limits_{k=1}^{K} a_k f[u_k(n)] \end{cases} \tag{4.14}$$

Original Hammerstein (OH) model [13] is achieved by reducing the multi-dimensional Volterra kernel to one-dimensional kernel as Eq. 4.15, which means only when the index $m_1, m_2, \ldots, m_p$ are equal to each other, then kernel has non-zero values.

$$h_p(m_1, m_2, \ldots, m_p) = a_p h(m, m, \ldots, m) = a_p h(m) \tag{4.15}$$

And the OH model can be written as Eq. 4.16.

$$\begin{cases} u(n) = \sum\limits_{p=1}^{P} a_p [x(n)]^p \\ y_{\mathrm{OH}}(n) = \sum\limits_{m=0}^{M} h(m) u(n-m) \end{cases} \tag{4.16}$$

Similar as OW model, the OH model has a clear physical architecture, i.e., it is formed by a static P-order nonlinear power series function followed by a $(M + 1)$-tap FIR filter. In order to enhance the modelling performance, an Augmented Hammerstein (AH) model was presented [14] as below.

$$\begin{cases} u(n) = \sum\limits_{p=1}^{P} a_p [x(n)]^p \\ y_{\mathrm{AH}}(n) = \sum\limits_{m=0}^{M_1} h_1(m) u(n-m) + \sum\limits_{m=0}^{M_2} h_2(m) u(n-m) |u(n-m)| \end{cases} \tag{4.17}$$

Dynamic deviation reduction-based Volterra model (DDR model) was introduced in [15] to simplify the full Volterra series by truncating the Volterra series to the limited dynamics order. The 1st-order [16] and simplified 2nd-order DDR-based Volterra model [17] can be written as Eqs. 4.18 and 4.19, respectively.

$$\begin{aligned} y_{\mathrm{DDR}}(n) = & \sum_{p=0}^{\frac{P-1}{2}} \sum_{m=0}^{M} a_{pm} |x(n)|^{2p} x(n-m) \\ & + \sum_{p=1}^{\frac{P-1}{2}} \sum_{m=1}^{M} b_{pm} |x(n)|^{2(p-1)} x^2(n) x^*(n-m) \end{aligned} \tag{4.18}$$

$$\begin{aligned} y_{\mathrm{SDDR}}(n) = & \sum_{p=0}^{\frac{P-1}{2}} \sum_{m=0}^{M} a_{pm} |x(n)|^{2p} x(n-m) \\ & + \sum_{p=1}^{\frac{P-1}{2}} \sum_{m=1}^{M} b_{pm} |x(n)|^{2(p-1)} x^2(n) x^*(n-m) \\ & + \sum_{p=1}^{\frac{P-1}{2}} \sum_{m=1}^{M} c_{pm} |x(n)|^{2(p-1)} x(n) |x(n-m)|^2 \\ & + \sum_{p=1}^{\frac{P-1}{2}} \sum_{m=1}^{M} d_{pm} |x(n)|^{2(p-1)} x^*(n) x^2(n-m) \end{aligned} \tag{4.19}$$

where the a_{pm}, b_{pm}, c_{pm} and d_{pm} are the corresponding coefficients, and $(\cdot)^*$ represents the conjugate function.

Memory polynomial (MP) model [18] is obtained by firstly removing all of the cross terms generated by $x(n)$ and its delayed versions $x(n - m_k)$ in a full Volterra series and then keep the terms that have the same phase information as the original complex signal as below.

$$y_{MP}(n) = \sum_{p=1}^{P}\sum_{m=0}^{M} h_{pm}x(n-m)|x(n-m)|^{p-1} \tag{4.20}$$

Generalized memory polynomial (GMP) [19] was introduced to enhance the modelling capability of MP model by taking the local memory terms $x(n - m \pm l)$ that are close to the current memory term $x(n - m)$ into account. The GMP model can be mathematically represented as below.

$$\begin{aligned} y_{GMP}(n) = & \sum_{p=1}^{P_a}\sum_{m=0}^{M_a} a_{pm}x(n-m)|x(n-m)|^{p-1} \\ & + \sum_{p=2}^{P_b}\sum_{m=0}^{M_b}\sum_{l=1}^{L_b} b_{pml}x(n-m)|x(n-m-l)|^{p-1} \\ & + \sum_{p=2}^{P_c}\sum_{m=0}^{M_c}\sum_{l=1}^{L_c} c_{pml}x(n-m)|x(n-m+l)|^{p-1} \end{aligned} \tag{4.21}$$

After somewhat of simplification, the complexity of the nonlinear convolution models is reduced down to a "comfortable" level that can be handled by the physical state-of-the-art processors, like FPGA, for real-time applications. It is not difficult to realize that the basic common computation units of those models are multiply-add unit (memory-associated) and multiply-multiply unit (nonlinearity-associated). The simple mapping functions, like from $x(n)$ to a constellation set $\{x(n), -x(n), x^*(n), -x^*(n)\}$, can be easily achieved by manipulating the sign bit of the complex signal from FPGA point view.

$$\begin{cases} x(n) = +x_I + jx_Q \\ -x(n) = -x_I - jx_Q \\ x^*(n) = +x_I - jx_Q \\ -x^*(n) = -x_I + jx_Q \end{cases} \tag{4.22}$$

4.3.2 Direct Synthesizable Architecture

We will use memory polynomial model as an example to illustrate the FPGA implementation architecture. Given the nonlinear order $P = 5$ and memory length $M = 2$, the MP model can be written as,

$$
\begin{aligned}
y_{\mathrm{MP}}(n) &= \sum_{p=1}^{5}\sum_{m=0}^{2} h_{pm}x(n-m)|x(n-m)|^{p-1} \\
&= h_{10}x(n) + h_{11}x(n-1) + h_{12}x(n-2) \\
&\quad + h_{20}x(n)|x(n)| + h_{21}x(n-1)|x(n-1)| + h_{22}x(n-2)|x(n-2)| \\
&\quad + h_{30}x(n)|x(n)|^2 + h_{31}x(n-1)|x(n-1)|^2 + h_{32}x(n-2)|x(n-2)|^2 \\
&\quad + h_{40}x(n)|x(n)|^3 + h_{41}x(n-1)|x(n-1)|^3 + h_{42}x(n-2)|x(n-2)|^3 \\
&\quad + h_{50}x(n)|x(n)|^4 + h_{51}x(n-1)|x(n-1)|^4 + h_{52}x(n-2)|x(n-2)|^4
\end{aligned}
\tag{4.23}
$$

The direct synthesizable FPGA implementation architecture is simple, i.e., using complex multipliers to perform every single complex multiplication operation required including the multiplications between power terms, and multiplications between coefficients and terms. Then total number of complex multipliers would be $(1 + 2 + 3 + 4 + 5) \times 3 = 45$. However, if we reuse some of the intermediate values and share some common architecture, we will reach to a simplified direct synthesizable FPGA implementation architecture as shown in Fig. 4.10, where the symbol $\otimes$ represents the complex multiplication operation.

The symbol z^{-n_1} and z^{-n_2} represent the latency of the module $|\cdot|$ and complex multiplication operation $\otimes$, respectively, which need to be taken into account when performing aggregations in the real FPGA hardware. This is because in order to carry out a proper function, the signals at the same stage in different parallel processing branches need to be synchronized in the time domain before aggregating them together. This simplified direct implementation architecture is straightforward but still quite a few hardware multipliers are required. In the example, the number of utilized complex multipliers is 19, which is corresponding to 57 hardware multipliers (e.g., DSP48s on the Xilinx FPGAs, and we used the compact 3-hardware multiplier architecture to construct complex multiplication). The absolute value of the input signal $|x(n)|$ can be derived efficiently by using coordinate rotational digital computer (CORDIC) algorithm on a FPGA, which actually can perform square root function. The detailed explanation and application case of CORDIC can be found online [20] from FPGA vendor Xilinx.

The processing latency from the first input to the first output T_{DS} can be derived in the following equation as a general case.

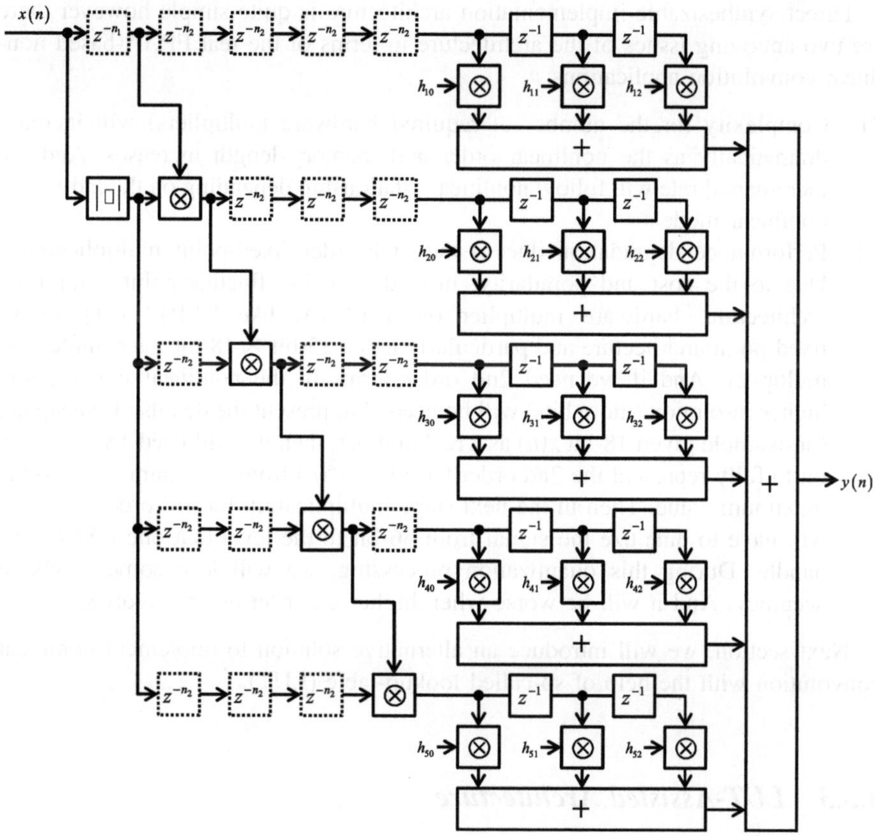

Fig. 4.10 Simplified direct synthesizable implementation architecture of MP model

$$T_{DS} = T_{|\cdot|} + (P-1)T_{CM} + T_{LC} + T_{Agg} \quad (4.24)$$

The symbol $T_{|\cdot|}$, T_{CM}, T_{LC} and T_{Agg} represent the processing latency (clock cycles) for absolute processing stage, complex multiplication stage, linear convolution stage (for memory part) and aggregation stage, respectively. The T_{LC} depends on the actual architecture in place to realize the memory part (detailed explanation is in Chap. 3), and T_{Agg} can be achieved by either addition-chain architecture or addition-tree architecture with the following latencies. T_A represents the latency for complex addition operation.

$$T_{Agg} = (P-1)T_A \quad \text{(addition-chain)} \quad (4.25)$$

$$T_{Agg} = \lceil \log_2 P \rceil_c T_A \quad \text{(addition-tree)} \quad (4.26)$$

Direct synthesizable implementation architecture is quite simple however there are two annoying issues of the architecture in terms of the real FPGA-based nonlinear convolution applications:

(1) Complexity (or the number of required hardware multipliers) will increase dramatically as the nonlinear order and memory length increases. And the incremental rate will follow nonlinear relationship depending on the individual nonlinear models.
(2) Performance degradation due to the high-order fixed-point multiplications. Due to the cost and popularity, instead of using floating-point computing architecture, hardware multiplier on a FPGA, like DSP48 [21] adopts fixed-point architecture and particularly it is a 25-bit × 18-bit 2's complement multiplier. And if we need 2nd order terms in the nonlinear convolution, higher resolution (more bits) will be needed to present the details of the signal, for example given 18-bit $x(n)$ as a real number, then we will need $18 \times 2 = 36$ bit to fully represent the 2nd order term like $x^2(n)$ from minimum value to the maximum value. Then in the next stage multiplication for 3rd order term, we will have to quantize the signal from 36-bit to the level that one DSP48 can handle. During this quantization processing, we will lose some levels of accuracy. And it will be worse when higher order terms are involved.

Next section, we will introduce an alternative solution to implement nonlinear convolution with the help of so-called lookup-table (LUT).

4.3.3 LUT-Assisted Architecture

From Eq. 4.20, we can see that the number of the multipliers required by MP model mainly results from the generation of the power and memory terms, and their multiplications with the corresponding coefficients. However, if we rewrite the MP model in the following manner,

$$\begin{aligned} y_{\mathrm{MP}}(n) &= \sum_{m=0}^{M} g_m x(n-m) \\ where\ g_m &= \sum_{p=1}^{P} h_{pm} |x(n-m)|^{p-1} \end{aligned} \tag{4.27}$$

It is not difficult to find that, given a memory tap m, g_m can be considered as a constant complex gain to memory term $x(n-m)$ and g_m depends on polynomial power terms of $|x(n)|$ only, since in a static system or segmental static system (switching-based dynamic system) the coefficients h_{pm} can be considered as constant during a period or in time-invariant operations when m is fixed. In a real application, like wireless communication, even if the instantaneous samples of the

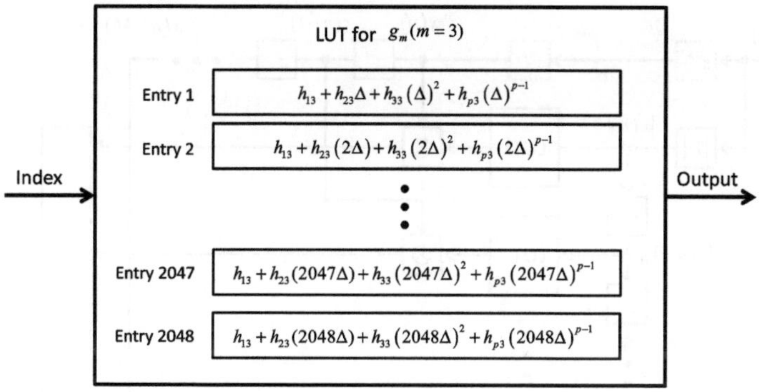

Fig. 4.11 An example of LUT built for g_3

signal $x(n)$ and $x(n - m)$ are randomly created, the range of the signal is normally known in advance. Therefore, it is possible to calculate the high-order power terms of $|x(n)|$, i.e., g_m in advance and store them in a table and look-up them later. This approach is named as look-up table (LUT)—assisted gain indexing [22]. For example, if we normalize the maximum input signal value to unity and use a LUT of size 2048, then the value of each LUT entry can be calculated according to Eq. 4.27. Figure 4.11 illustrates an example of LUT contents for g_3.

In this way, we can build $M + 1$ LUTs helping to achieve MP model instead of using a bulk of multipliers, and the number of LUTs doesn't increase with nonlinear order P. Although during the calculation of g_m high precision signal processing is required to produce accurate results, e.g., producing high-order power terms, this does not affect the final LUT structure because the intermediate values can be calculated offline in a software environment and they are not stored in the LUTs, or they can be calculated on-line in a System on-Chip (SoC), which is sitting either inside the FPGA chip, like Xilinx Zynq SoC or Intel Stratix 10 SoC, or outside the FPGA chip with certain inter-connections. Since only the final results of g_m are stored in the LUTs, the precision requirement for these final values is much lower than that required for the intermediate power terms. And therefore, they can be easily scaled to suit a small LUT size significantly saving on-chip memory storage or a large LUT size boosting the processing accuracy. An example of using LUT-assisted approach to implement MP model is illustrated in Fig. 4.12, where the symbol z^{-n_1} and z^{-n_3} represent the latency of absolute module and LUT module, respectively. The number of required complex multipliers is significantly reduced down to $(M + 1)$ depending on the length of memory.

Furthermore, if we take the actual processing rate to data rate ratio into account, we can possibly further reduce the complexity in a similar TDM approach introduced in the linear convolution chapter. For example, if $(M + 1) \times CLK_{DATA} < CLK_{FPGA}$, we can achieve $(M + 1)$ taps memory part with only one complex multiplier in a fully-serial linear convolution manner, and the outputs from each LUT can be considered as the coefficients for different taps. Figure 4.13 illustrates

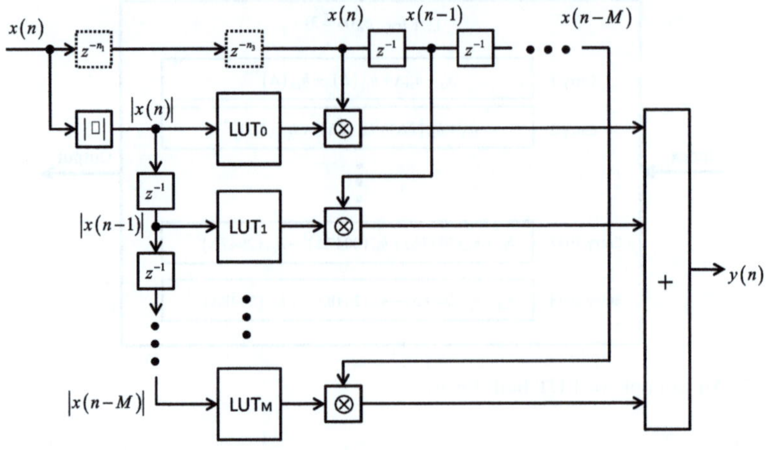

Fig. 4.12 An example of the LUT-assisted MP model implementation architecture

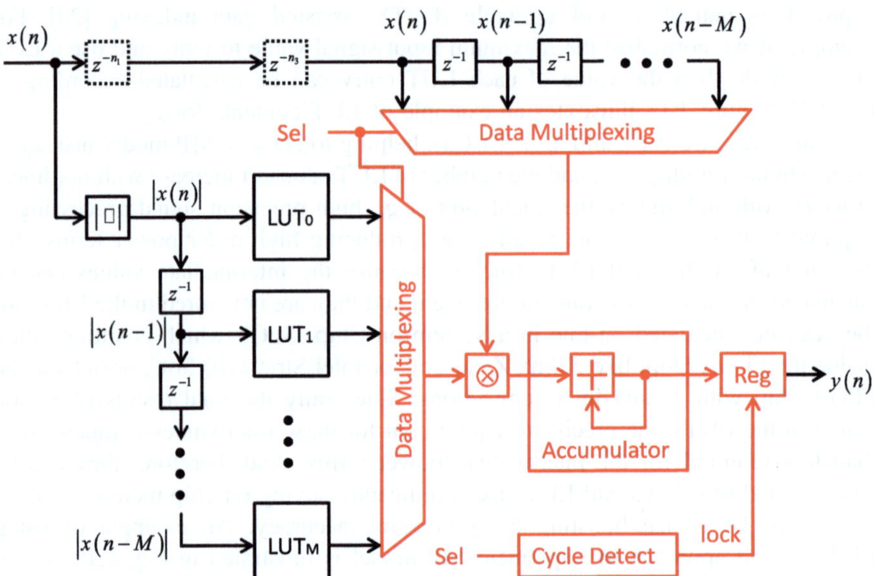

Fig. 4.13 An example of the LUT-assisted MP model implementation architecture combined with fully serial linear convolution for the memory part

an example implementation of MP model using LUT-assisted approach combined with fully-serial linear convolution approach (The detailed fully-serial architecture explanation can be found in Chap. 3). Semi-parallel architecture can be utilized as well according to the requirement of the system and the relationship between $(M + 1) \times CLK_{\mathrm{DATA}}$ and FPGA processing rate CLK_{FPGA}, for the sake of the simplicity, we will not provide an example of this, the essentials of the semi-parallel architecture can be found in Chap. 3.

Without the loss of generality, the processing latency from the first input to the first output T_{LA} can be derived in the following equation.

$$T_{LA} = T_{|\cdot|} + T_{LUT} + T_{LC} \tag{4.28}$$

The symbol $T_{|\cdot|}$, T_{LUT} and T_{LC} represent the processing latency (clock cycles) for absolute processing stage, LUT stage and linear convolution stage (for memory part), respectively. The T_{LUT} usually will take just a few processing clocks due to the dual-port memory read operation latency. The T_{LC} depends on the actual architecture (e.g., either fully-parallel architecture, or semi-parallel architecture or fully-serial architecture) in place to realize the memory part (detailed explanation is in Chap. 3).

4.3.4 Architecture Comparison

In general, for low-order polynomial-type nonlinear convolution applications, the straightforward direct synthesizable implementation architecture can be adopted as the first trial approach to evaluate the architecture complexity against the system requirement in terms of the resource utilization and quantization error tolerance. While for high-order nonlinear system scenario, LUT-assisted architecture is recommended as an efficient and alternative approach to perform nonlinear convolution functions. The latter approach actually is an integration approach, which only produces the necessary and useful outputs via a black-box type mapping function (LUT). The construction of the LUT contents can be a static (off-line processing) or dynamic (on-line processing) approach.

Depending on the system requirement, we should adopt suitable implementation architecture for performing nonlinear convolution on a FPGA. Table 4.2 summaries the key differences between the two different implementation architecture when implementing the MP model. The other nonlinear models may have different

Table 4.2 FPGA-based nonlinear convolution implementation architecture comparison (using MP model as an example)

	Direct synthesizable approach	LUT-assisted approach
CLK_{FPGA}	CLK_{DATA}	$\alpha \cdot CLK_{\text{DATA}}^{1}$
Number of CM	$\frac{(P+1)P}{2} \cdot (M+1)$	$\frac{M+1}{\alpha}$
Processing latency	$T_{|\cdot|} + (P-1)T_{CM} + T_{LC} + T_{Agg}$	$T_{|\cdot|} + T_{LUT} + T_{LC}$

1. Scenario $\alpha = 1$ refers to the implementation architecture with fully-parallel architecture-based memory part, scenario $\alpha = M + 1$ refers to the implementation architecture with fully-serial architecture-based memory part and $1 < \alpha < M + 1$ refers to the implementation architecture with semi-parallel architecture-based memory part

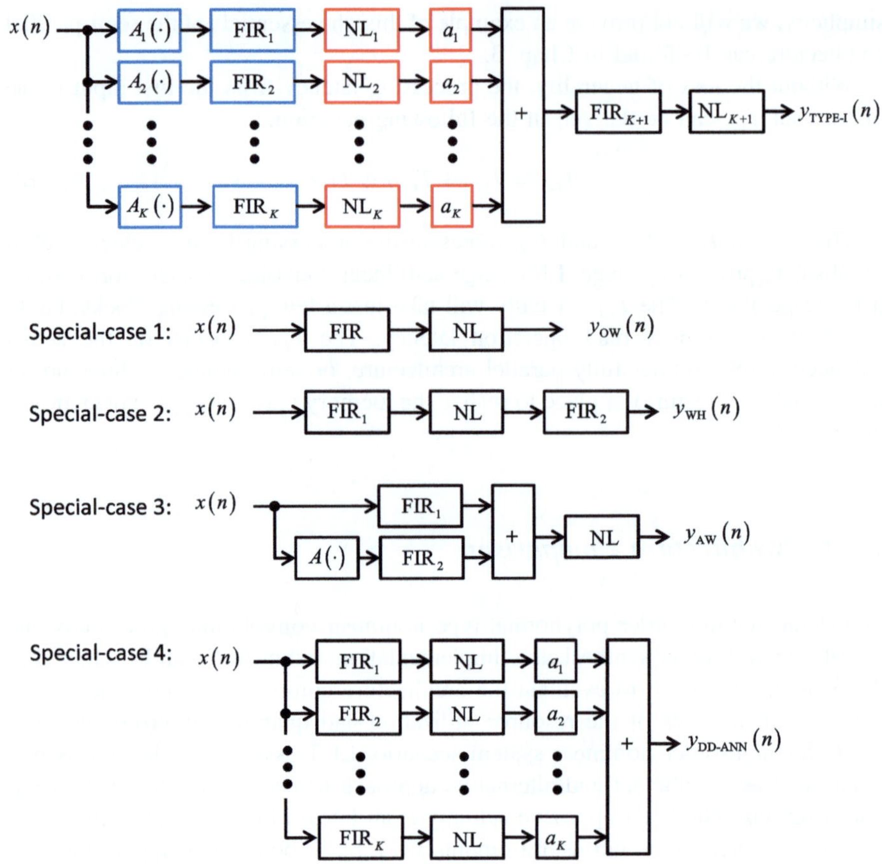

Fig. 4.14 Type-I generalized nonlinear convolution architecture and its special cases

parameters however the base architecture would be similar for all of those polynomial-type nonlinear convolutions.

In order to show the architectural-difference of individual nonlinear models, the following Figs. 4.14 and 4.15) illustrate the abstracted generalized implementation diagrams of the popular nonlinear models. We started from two generalized architectures in which either memory function occurs before nonlinear function or nonlinear function occurs before memory function. The two generalized architectures are constructed by a few basic common functions including nonlinear function (NL_k), branch scaling function (a_k), augment function ($A_k(\bullet)$) and memory function (FIR_k). The popular utilized nonlinear models can be considered as special cases of generalized nonlinear convolution architectures. We provide the examples of those special cases with corresponding parameter settings as shown in Tables 4.3 and 4.4.

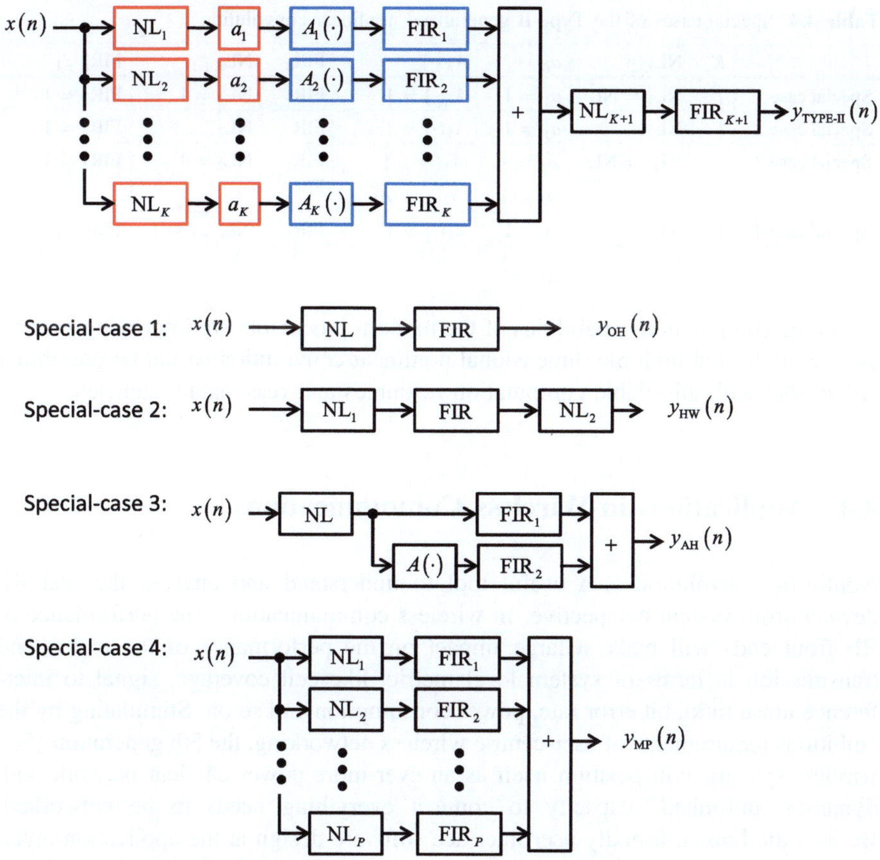

Fig. 4.15 Type-II generalized nonlinear convolution architecture and its special cases

Table 4.3 Special cases of the Type-I generalized nonlinear convolution

	K	$A_K(\cdot)$	FIR_K	NL_K	a_K	FIR_{K+1}	NL_{K+1}
Special case 1	1	$A_1(\cdot)=1$	FIR	$\text{NL}_1=\text{NL}$	$a_1=1$	$\text{FIR}_2=1$	$\text{NL}_2=1$
Special case 2	1	$A_1(\cdot)=1$	FIR_1	$\text{NL}_1=\text{NL}$	$a_1=1$	FIR_2	$\text{NL}_2=1$
Special case 3	2	$A_1(\cdot)=1$ $A_2(\cdot)=A(\cdot)$	FIR_1 FIR_2	$\text{NL}_1=1$ $\text{NL}_2=1$	$a_1=1$ $a_2=1$	$\text{FIR}_3=1$	$\text{NL}_3=\text{NL}$
Special case 4	K	$A_k(\cdot)=1$	FIR_k	$\text{NL}_k=\text{NL}$	$a_k=1$	$\text{FIR}_{K+1}=1$	$\text{NL}_{K+1}=1$

Table 4.4 Special cases of the Type-II generalized nonlinear convolution

	K	NL_K	a_K	$A_K(\cdot)$	FIR_K	NL_{K+1}	FIR_{K+1}
Special case 1	1	$NL_1 = NL$	$a_1 = 1$	$A_1(\cdot) = 1$	FIR	$NL_2 = 1$	$FIR_2 = 1$
Special case 2	1	NL_1	$a_1 = 1$	$A_1(\cdot) = 1$	FIR	NL_2	$FIR_2 = 1$
Special case 3	2	$NL_1 = NL$	$a_1 = 1$ $a_2 = 1$	$A_1(\cdot) = 1$ $A_2(\cdot) = A(\cdot)$	FIR_1 FIR_2	$NL_2 = 1$	$FIR_3 = 1$
Special case 4	P	NL_k	$a_k = 1$	$A_k(\cdot) = 1$	FIR_k	$NL_{K+1} = 1$	$FIR_{K+1} = 1$

As the computation capabilities of the modern processors are improving year by year, complicated multiple-dimensional nonlinear convolution would be possibility carried out with affordable computation resources and reasonable latencies.

4.4 Applications in Wireless Communications

Nonlinear convolution is a useful tool to understand and analyze the real RF devices from system perspective. In wireless communications, the performance of RF front ends will make a large impact on the performance of the end-to-end transmission in terms of system level metric, like cell coverage, signal to interference noise ratio, bit error rate, power consumption and so on. Stimulating by the ambitious requirements of user-centric wireless networking, the 5th generation (5G) wireless systems will position itself as an ever-more power efficient network with dynamic "unlimited" capacity to connect everything needs to be networked. Besides the human-friendly user interface software design at the application layer, the user experience will be largely determined by the physical layer hardware performance of the mobile network. Simply speaking, 1 Mb/s network speed will not provide you high definition mobile video on the go. The fundamental game is how to effectively transmit a large amount of data over the wireless radio link.

Within the mobile network, comparing to the subscriber elements (mobile terminals) or network control elements, the costliest elements of the mobile network, i.e., base stations also consume the majority of the power feed to the whole network. In the base stations, the radio including power amplifiers and feeder will consume the dominant power, however the majority of the power used to transmit the signal is wasted in the form of the heat due to the relatively low power conversion efficiency of the practical RF power amplifiers [23].

In this section, we will start from the modelling of the nonlinear practical RF devices, particularly the power amplifiers, and then we will introduce the nonlinear convolution based digital predistortion (DPD) technique, which has been largely utilized in the efficient RF front ends using software and hardware co-operated nonlinear compensation approaches.

4.4.1 Nonlinear Modelling of RF Power Amplifiers

As one of the most expensive components in the RF front end, the power amplifier (PA) is a physical analog circuit for converting low power RF signal into a higher power RF signal. In order to achieve high power conversion efficiency, the PAs are usually required to be operated in peak power region or called saturation region. However, driving the PAs into saturations will introduce nonlinear distortions, which causes in-band distortion and out-of-band spectral regrowth [2, 3]. On one hand, the in-band distortion will damage the quality of the transmitted signal, consequently degrades the overall communication performance within the transmission frequency band. On the other hand, the out-of-band spectral regrowth turns into undesired interference to the free space, and this will degrade the transmission performance or even cause failure of any other wireless systems, which are operating at the adjacent frequency bands.

For short, the distortions introduced by PAs can be generalized categorized as follows: (1) memoryless nonlinearities, inherently induced by the nonlinearities of the physical components; (2) linear memory effects, arising from time delays or phase shift in the matching networks of device used; (3) nonlinear memory effects, caused by trapping effects, non-ideal bias networks, etc. In order to characterize the PA, RF engineers has been using circuit level modelling for a long time, which describes the physical nonlinear behaviour using equivalent circuits. Unlike the circuit level modelling, the black-box based behavioural modelling is carried out at the system level. This behavioural modelling uses a series of mathematical formulations (e.g., nonlinear convolution) for derive the relationship between input and output of the PA. The advantage of such black-box modelling is that, the complicated knowledge of the physical RF circuits is not required, thus the nonlinear characterization procedure tends to be much easier [2, 3], and moreover, once validated, those models can be directly utilized as the "inversed" nonlinear function to provide some digital predistortion pre-compensating for the nonlinear distortions and memory effects introduced by the RF PAs.

A typical behavioural modeling based PA performance evaluation platform is shown in Fig. 4.16.

This PA evaluation system includes the following essential units: (1) a PC with some signal processing and interface control software (e.g., Matlab), which generates a digital pattern or baseband signal to stimulate the transmission RF chain at the data forward path and captures the baseband signal at the data feedback path; (2) digital to RF chain for up-converting the signal from baseband to the device operation RF band. This unit typically contains digital to analog convertor (DAC), mixer (⊗), low pass filter (LPF), local oscillator (LO) and bass-pass filter (BPF); (3) device under test (DUT), i.e., RF PA with a coupler to couple a small amount of power (e.g., 20 dB attenuation) to the feedback path, which contains the same RF information; (4) RF to digital chain for down-converting the signal from RF to the baseband, which contains BPF, mixer, LO and analog to digital convertor (ADC).

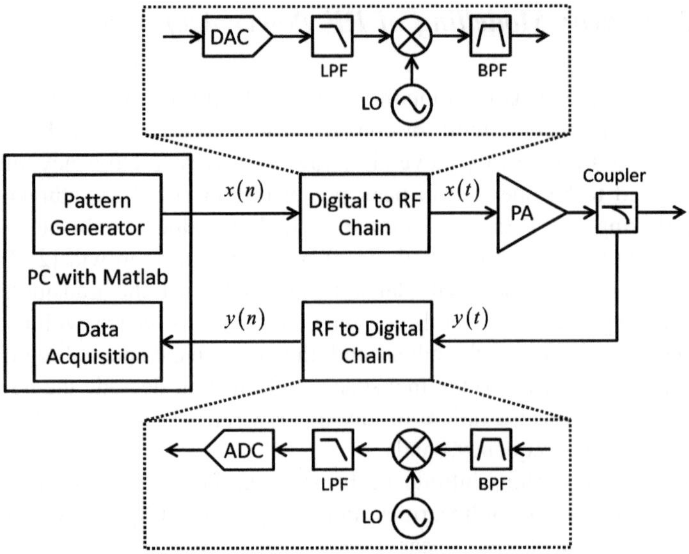

Fig. 4.16 Simplified diagram of a typical PA performance evaluation platform

Fig. 4.17 System diagram of PA modelling and verification

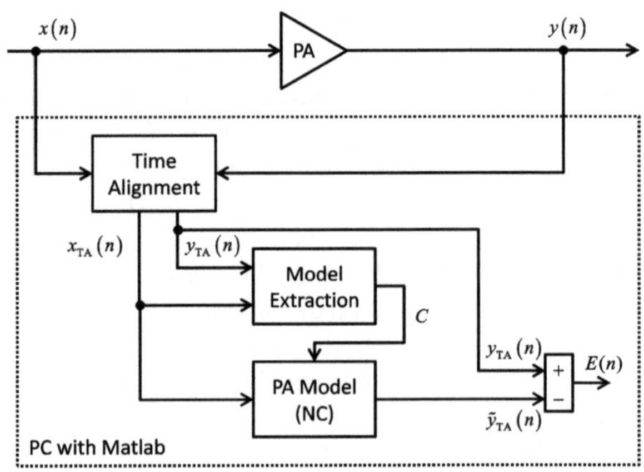

Figure 4.17 is a simplified diagram illustrating of PA behavioural modelling and model verification. And the typical instrument-based PA testing procedures are provided as below.

(1) Platform setup: at first we set up the PA evaluation platform with commercial instruments using signal generator (e.g., Keysight ESG4438c [24] or R&S SMA100a [25]) as digital to RF chain and spectrum analyzer (e.g., Keysight vector analyzer [26] or R&S FSQ26 [27]) as RF to digital chain, together with Matlab equipped PC.

(2) Data generation: we create a baseband signal waveform (a digital complex data sequence) at particular sampling rate, e.g., a 10 ms 4C-UMTS signal at 92.16 MSPS sampling rate, i.e., 921600 complex samples.
(3) PA excitation: we download the generated waveform to the signal generator and up-convert the baseband signal to the RF frequency, e.g., 2.14 GHz, to periodically excite the device under test.
(4) Data acquisition: we connect the PA output with a coupler and get back the RF signal to spectrum analyzer, which down-converts the RF signal to the complex baseband signal and captures a period of baseband signal. For the sake of simplicity, we usually capture the same length of transmitted signal, e.g., 921600 complex samples.
(5) Time alignment: if the capture procedure is not trigged by certain indicator showing the first transmitted sample, then the feedback signal sequence may not be properly aligned with transmitted signal in the time domain. Certain time alignment function (e.g., correlation-based rotation approach using convolution to find the most correlated peak and derive the corresponding delays between two sequences) needs to be carried out, so that we will have two time-aligned data sequences corresponding to the input and output of the PA, respectively.
(6) PA model extraction: given a nonlinear model with certain unknown coefficients, we can use small portion of the paired input and output PA data sequences (e.g., 8000 samples) to perform coefficient estimation function using certain estimator, like least squares (LS) estimation.
(7) Model validation: given the extracted model coefficients, we stimulate the black-box model with original input data sequence (nonlinear convolution processing) in a cross-validation manner, which uses small portion of the sequence to derive the modelling coefficients and applies the coefficients to the other section of the sequence. And we compare the output of black-box model and captured output of the PA. We will see very little difference or errors between model outputs and captured PA outputs if the model is suitable and properly extracted with certain coefficients.

In the time domain, we compare the PA output and model output in terms of normalized magnitude and phase as shown in Fig. 4.18, which verifies that the model output is very close to the actual PA output, i.e., the model can describe the real behaviour of the physical devices. On the contrary, certain degree differences will be seen if the model cannot characterize the device.

In the frequency domain, we can plot the PSD of the error signal and compare to the spectrum of the PA output. As shown in Fig. 4.19, an accurate model will lead to a small error in the frequency domain across the signal bandwidth. And less accurate model will generate certain degree errors in the frequency domain.

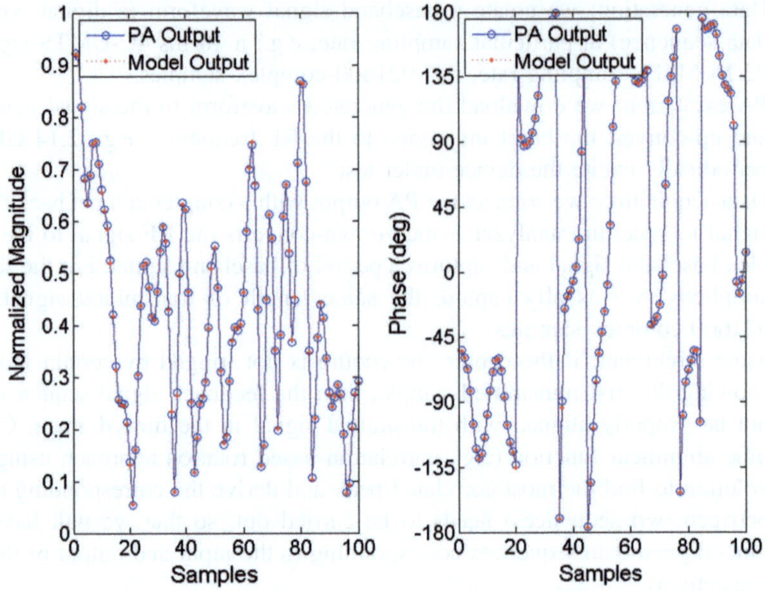

Fig. 4.18 Nonlinear modelling performance illustration in the time domain

Fig. 4.19 Nonlinear modelling performance illustration in the frequency domain

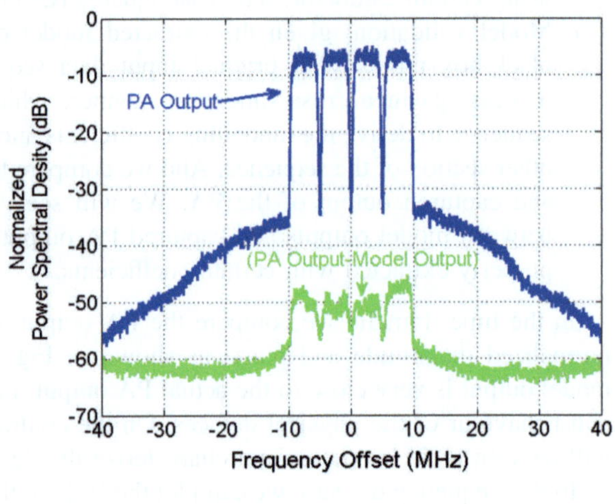

4.4.2 Digital Predistortion in Modern RF Front Ends

Digital predistortion (DPD) technique [3] has been largely utilized on the modern wireless transceivers for improving the linearity performance of saturated high-efficient PA. A simplified diagram of the a typical DPD-enabled wireless transmitter is shown in Fig. 4.20, which includes the components on the normal transmitter and an extra feedback receiver (or called sampling receiver), one digital predistorter unit and one coefficients calculation unit.

In general, the digital predistorter unit at the forward data path (normal data path) needs to be real-time (RT) due to the requirement of processing continuous incoming data streams, while the coefficient calculation function at the feedback path can be carried out in a non-real-time (NRT) way for periodical coefficients updating. The basic concept of DPD is quite simple that if we put in place a nonlinear function in front of another nonlinear function along the signal chain with proper "inversed" transfer function, the cascade system can provide a pseudo-linear transfer function regarding the input to output relationship as shown in Fig. 4.21. Usually in the real system, it is quite difficult to derive an exact closed-form mathematical equations of the inversed nonlinear PA behaviour with memory effects, instead some statistical estimation algorithms are quite popular and widely utilized, e.g., least squares (LS) [28, 29], to estimate the inversed behaviour minimizing the so-called mean square error (MSE) given certain constraints.

Without the loss of generality, DPD system can be considered as a switching-based dynamic nonlinear convolution processing system, and during each of the stable state a set of coefficients will be applied periodically to perform

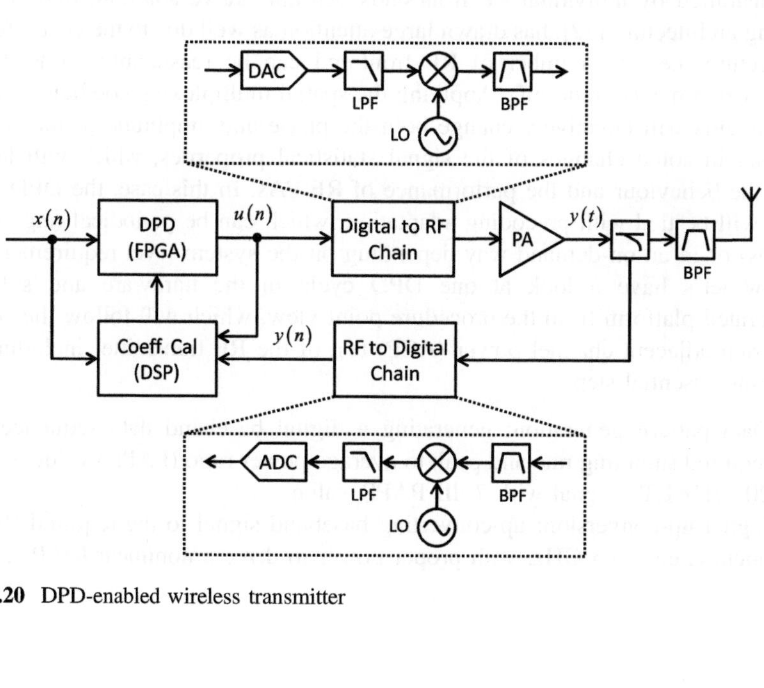

Fig. 4.20 DPD-enabled wireless transmitter

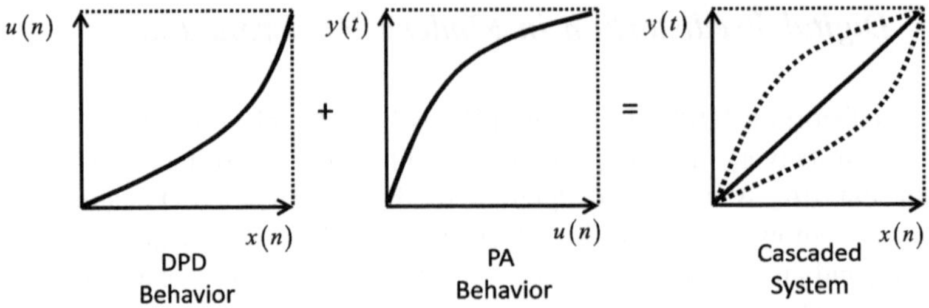

Fig. 4.21 DPD concept illustration

the inverse transfer function against the nonlinear behaviour of RF PAs. Depending on the switching transition time, DPD system can be categorized into three groups differentiated by the individual coefficients adaptive approach: (1) fast-periodical-adaptive DPD approaches, which provide fast DPD coefficients updating for the non-stationary bursty transmission scenario; (2) slow-periodical-adaptive DPD approaches, which will update the DPD coefficients for the periodical stationary transmission scenario. (3) on-demand-adaptive DPD approaches, which will update the DPD coefficients when it is necessary. However, in the real wireless application point view, the nonlinear convolution based DPD approach may require all of the above three coefficient adaptation methods. For example, in the future-proof massive MIMO system [30], frequency dependent spatial multiplexing technique will apply certain digital weighting coefficients generating multiple precoded data streams to be transmitted by the RF front-ends (here we use full digital precoding/beam forming as an example [31], where each precoded data stream will be transmitted by individual RF front-ends. An alternative solution, hybrid beam forming architecture [32], has drawn large attention as well due to the cost-effective architecture, i.e., the number of RF front-end can be reasonably reduced with certain performance trade-off). Applying the spatial multiplexing coefficients to the data streams will effectively change both the phase and amplitude of the signals, resulting in some changes of the signal statistical properties, which will largely affect the behaviour and the performance of RF PAs. In this case, the DPD adaptation will be tied with precoding adaptation, which can be periodical (e.g., every 100 ms) or in an on-demand way depending on the system level requirement.

Now let's have a look at one DPD cycle in the hardware and software co-operated platform from the procedure point view, which will follow the similar multi-step adjacent channel power (ACP) test of the RF transmitter including the following essential steps:

(1) Data pattern generation: generating a digital baseband data sequence with required sampling rate and peak to average power ratio (PAPR) value, such as 20 MHz LTE signal with 7 dB PAPR value.
(2) Signal up-conversion: up-converting baseband signal to the required RF frequency, e.g., 3.5 GHz, with proper power to drive a nonlinear RF PA.

(3) Feedback signal acquisition: down-converting the output of the RF PA to the baseband or intermediate frequency (IF) and then capturing the baseband or IF analog signal with proper ADC-based digitizer.
(4) Time alignment: aligning the original input and captured PA output in the time domain so that the latency introduced by the system (transmitter and feedback path) can be characterized and compensated, thus obtaining matched/paired PA input and output in the digital domain.
(5) DPD parameter extraction and updating: calculating DPD parameters (nonlinear convolution coefficients) and then updating the parameters in the nonlinear convolution-like digital predistorter sitting in the processors, like FPGA.
(6) Performance assessment: using spectrum analyser to evaluate the frequency-domain linearization performance by measuring the adjacent channel leakage power (ACLR) or by capturing the signal again in the digital domain to assess the time-domain performance, like error vector magnitude (EVM) or normalized mean square error (NMSE).

As the evolutionary development of modern FPGA, a single chip DPD solution is quite popular for wireless applications. Figure 4.22 illustrates a simplified diagram of FPGA-based single-chip DPD solution, which performs the DPD training and nonlinear convolution (Digital Predistorter) by the same FPGA chip on the modern wireless transmitter. The forward DPD data path is carried out by FPGA fabric while the DPD training path is performed by e.g., either hardware-core ARM [33] or software-core Microblaze (MB) [34] on a Xilinx FPGA or equivalent software processor Nios II [35] on an Intel (Altera) FPGA. For maximizing the DPD performance, in the Fig. 4.22, we put generalized dual-loop DPD training architecture in place [36], which can employ either so-called indirect learning

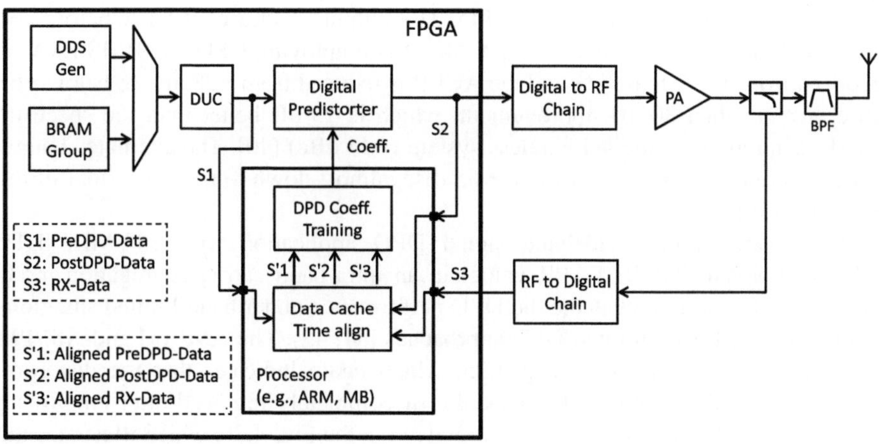

Fig. 4.22 Simplified diagram of the FPGA-based single-chip DPD solution

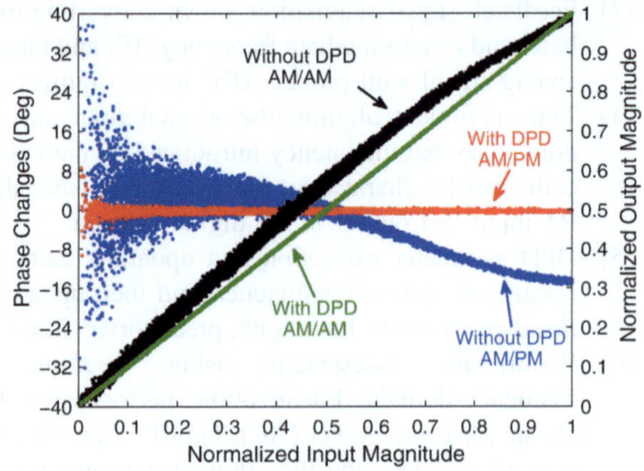

Fig. 4.23 AM/AM and AM/PM plots for the PA outputs with and without DPD, using 20 MHz LTE signal as excitation (© Cambridge University Press and the European Microwave Association [39], Reprinted with permission)

(using S2 and S3 to train DPD coefficients, like the ones used in [16, 19]) or direct learning (using S1 and S3 to train DPD coefficients [37, 38]) or dual-loop joint learning (using S1, S2 and S3 to train DPD coefficients [36]).

Figures 4.23 and 4.24 illustrate a typical DPD performance in the time domain and frequency domain using 20 MHz LTE signal as excitation [39]. A Doherty RF PA was excited by a LTE 20 MHz signal with 6.5 dB PAPR, and simplified second-order DDR-based Volterra model was implemented by the LUT-assisted nonlinear convolution approach. The AM/AM and AM/PM characteristics are given in the Fig. 4.23, which indicates that nonlinearities and memory effects are almost completely removed after DPD. The output spectra of the PA before and after DPD are also given in the Fig. 4.24. After employing DPD, the RMS EVM is reduced from 12.92 to 0.38% and the ACLR is reduced from −28 dBc to below −61 dBc, a more than 30 dB improvement, which is 15 dB better than the spectrum mask requirement in the 4G wireless system (−45 dBc) [40]. The alternate channel ACLR is reduced from −48 dBc to −65 dBc, almost down to the noise floor of the platform [39].

Moreover, an ultra-wideband signal DPD application with 100 MHz LTE advanced signal (7.8 dB PAPR value) in an intra-band carrier-aggregation mode was utilized as well to evaluate the DPD performance under band-limited situations using advanced band-limited DPD approaches [41, 42]. The AM/AM and AM/PM characteristics are shown in Fig. 4.25, which again indicates that the distortion introduced by the nonlinear PA was almost completely removed after employing DPD module. The frequency spectra are given in the Fig. 4.26, which illustrates the ACLR performance improvement from −28 dBc to below −50 dBc within the bandwidth regulated by the designed platform.

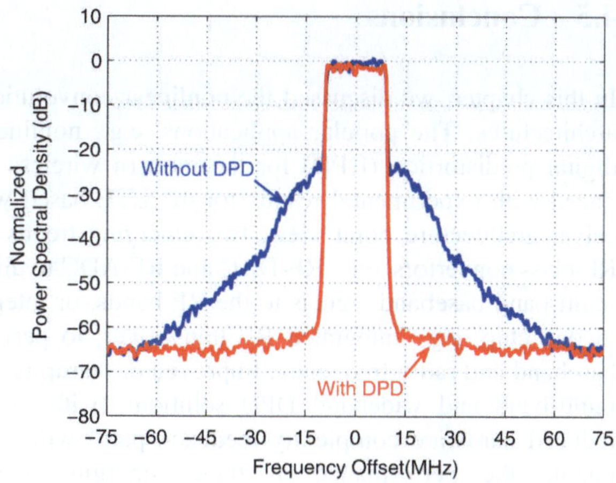

Fig. 4.24 Output spectra for the PA outputs with and without DPD, using 20 MHz LTE signal as excitation (© Cambridge University Press and the European Microwave Association [39], Reprinted with permission)

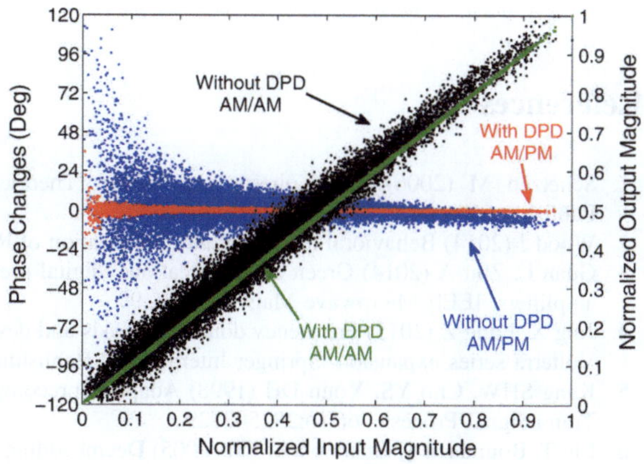

Fig. 4.25 AM/AM and AM/PM plots for the PA outputs with and without DPD, using 100 MHz LTE advanced signal as excitation (© Cambridge University Press and the European Microwave Association [39], Reprinted with permission)

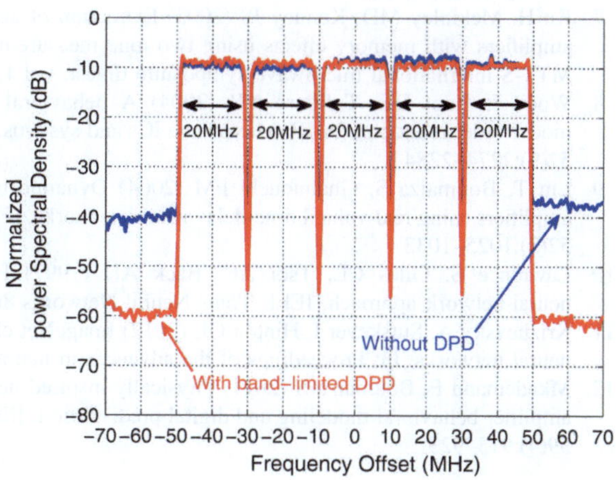

Fig. 4.26 Output spectra for the PA outputs with and without DPD, using 100 MHz LTE advanced signal as excitation (© Cambridge University Press and the European Microwave Association [39], Reprinted with permission)

4.5 Conclusions

In this chapter, we discussed the nonlinear convolution basis and its FPGA-based architectures. The popular applications, e.g., nonlinear RF device modeling and digital predistortion (DPD) for the modern wireless transceiver, were also introduced with experimental results for 4G LTE based wireless systems. From transceiver architecture point view, two emerging trends are rapidly developed using RF-class convertors, e.g., RF-DAC and RF-ADC to directly synthesize wideband or multi-band baseband signals to the RF bands, or integrating more RF components in one chip, e.g., integrated RF transceiver, to perform the conversion between baseband and radio frequency. Improved or disruptive nonlinear convolution-based multi-band and wideband DPD solutions (with lower power consumption and reduced hardware complexity feedback path) will be on high demand for maximizing the performance of those emerging RF systems in the wireless communications.

References

1. Schetzen M (2006) The Volterra and Wiener Theories of nonlinear systems. Krieger Publishing Company, Revised version
2. Wood J (2014) Behavioral modeling and linearization of RF power amplifiers. Artech House
3. Guan L, Zhu A (2014) Green communications: digital predistortion for wideband RF power amplifiers. IEEE Microwave Mag 15(7):84–99
4. Jing X, Lang Z (2015) Frequency domain analysis and design of nonlinear systems based on Volterra series expansion. Springer International Publishing, Berlin
5. Kang SHW, Cho YS, Youn DH (1998) Adaptive precompensation of Wiener system. IEEE Trans Signal Process 46(10):2825–2829
6. Liu T, Boumaiza S, Ghannouchi FM (2005) Deembedding static nonlinearities and accurately identifying and modeling memory effects in wide-band RF transmitters. IEE Trans Microw Theory Tech 53(11):3578–3587
7. Ku H, Mckinley MD, Kenney JS (2002) Extraction of accurate behavior models for power amplifiers with memory effects using two tone measurements. In: Proceedings of the IEEE MTT-S international microwave symposium digest, vol 1, pp 139–142
8. Wood J, Root DE, Tufillaro NB (2004) A behavioral modeling approach to nonlinear model-order reduction for RF/Microwave ICs and systems. IEEE Trans Microw Theory Tech 52(9):2274–2284
9. Liu T, Boumaiza S, Ghannouchi FM (2004) Dynamic behavioral modeling of 3G power amplifiers using real-valued time-delay neural networks. IEEE Trans Microwave Theory Tech 52(3):1025–1033
10. Lawrence S, Giles CL, Tsoi AC, Back AD (1997) Face recognition: a convolutional neural-network approach. IEEE Trans Neural Networks 8(1):98–113
11. Krizhevsky A, Sutskever I, Hinton GE (2012) ImageNet classification with deep convolution neural networks. In: Proceedings of the advances in neural information processing systems
12. Mkademand F, Boumaiza S (2011) Physically inspired neural network model for RF power amplifier behavioral modeling and digital predistortion. IEEE Trans Microwave Theory Tech 59(4):913–923

13. Gilabert PL, Montoro G, Bertran E (2005) On the Wiener and Hammerstein models for power amplifier predistortion. Proc Asia-Pacific Microw Conf 2:1–3
14. Liu T, Boumaiza S, Ghannouchi FM (2006) Augmented Hammerstein predistorter for linearization of broad-band wireless transmitters. IEEE Trans Microw Theory Tech 54 (4):1340–1349
15. Zhu A, Pedro JC, Brazil TJ (2006) Dynamic deviation reduction based Volterra behavioral modeling of RF power amplifiers. IEEE Trans Microw Theory Tech 54(12):4323–4332
16. Zhu A, Draxler PJ, Yan JJ, Brazil TJ, Kimball DF, Asbeck PM (2008) Open-loop digital predistorter for RF power amplifiers using dynamic deviation reduction-based Volterra series. IEEE Trans Microw Theory Tech 56(7):1524–1534
17. Guan L, Zhu A (2011) Simplified dynamic deviation reduction based Volterra model for doherty power amplifiers. In: Proceedings of the workshop integrated nonlinear microwave millimetre-wave circuits, pp 1–4
18. Kim J, Konstantinou K (2001) Digital predistortion of wideband signals based on power amplifier model with memory. Electron Lett 37(23):1417–1418
19. Morgan DR, Ma Z, Kim J, Zierdt MG, Pastalan J (2006) A generalized memory polynomial model for digital predistortion of RF power amplifiers. IEEE Trans Signal Process 54 (10):3852–3860
20. Xilinx (2016) CORDIC v6.0. LogiCore IP Product Guide, PG105
21. Xilinx (2016) 7 Series DSP48E1 Slice. User Guide, UG479
22. Guan L, Zhu A (2010) Low-cost FPGA implementation of Volterra series-based digital predistorter for RF power amplifiers. IEEE Trans Microw Theory Tech 58(4):866–872
23. Guan L (2012) High performance digital predistortion for wideband RF power amplifier. Ph.D. dissertation, School of Electrical, Electronics, Communication Engineering, University College Dublin, Dublin, Ireland, Jan 2012
24. Keysight company (2014) E4438C ESG Vector Signal Generator. User Guide
25. R&S company (2016) SMA 100A Signal Generator. Operating Manual v14
26. Keysight company (2013) E4440A PSA Spectrum Analyzer. User Guide
27. R&S company (2014) FSQ26 Spectrum Analyzer. User Manual
28. Ding L, Ma Z, Morgan DR, Zierdt M, Pastalan J (2006) A least squares/Newton method for digital predistortion of wideband signals. IEEE Trans Commun 54(5):833–840
29. Guan L, Zhu A (2012) Optimized low-complexity implementation of least squares based model extraction for digital predistortion of RF power amplifiers. IEEE Trans Microw Theory Tech 60(3):594–603
30. Weldon MK (2015) The Future X Network: a Bell Labs Perspective. CRC Press
31. Guan L (2015) Compact scalable frequency dependent precoding and its FPGA implementation for 5G massive MIMO wireless systems. Electron Lett 51(23):1937–1939
32. Venkateswaran V, Pivit F, Guan L (2016) Hybrid RF and digital beam former for cellular networks: algorithms, microwave architectures and measurements. IEEE Trans Microw Theory Tech 64(7):2226–2243
33. Xilinx (2016) Unleash the unparalleled power and flexibility of Zynq UltraScale + MPSoC. White Paper, wp 470
34. Xilinx (2016) Using the microblaze processor to accelerate cost-sensitive embedded system development. White Paper, wp 469
35. Altera (an Intel company) (2016) Nios II classic processor reference guide. NII5v1
36. Guan L, Zhu A (2011) Dual-loop model extraction for digital predistortion of wideband RF power amplifiers. IEEE Microwave Compon Lett 21(9):501–503
37. Braithwaite RN, Carichner S (2009) An improved Doherty amplifier using cascaded digital predistortion and digital gate voltage enhancement. IEEE Trans Microw Theory Tech 57 (12):3118–3126
38. Ding L, Mujica F, Yang Z (2013) Digital predistortion using direct learning with reduced bandwidth feedback. IEEE MTT-S Int Microw Symp Dig, 1–3

39. Guan L, Kearney R, Yu C, Zhu A (2013) High-performance Digital Predistortion test platform development for wideband RF power amplifiers. Int J Microw Wireless Technol 5 (02):149–162
40. E-UTRA (2012) Radio Frequency System Scenarios, Release 11, 3GPP Standard TR 36.942 V11.0.0
41. Yu C, Guan L, Zhu E, Zhu A (2012) Band-limited Volterra series based digital predistortion for wideband RF power amplifiers. IEEE Trans Microw Theory Tech 60(12):4198–4208
42. Guan L, Yu C, Zhu A (2012) Bandwidth-constrained Least Squares-based model extraction for band-limited digital predistortion of RF power amplifiers. In: IEEE international workshop on "integrated nonlinear microwave and millimeterwave circuits, INMMIC", Dublin, Ireland

Chapter 5
Advanced FPGA-based Fast Linear Convolution

Abstract Multiple long-tap digital convolutions are sometimes required in different appealing wireless applications, like Massive MIMO system in the 5th generation wireless communications. However, the complexity of such processing system will increase significantly as the number of digital convolution modules is getting larger and the tap of each convolution is growing longer. This chapter will focus on low complexity fast linear convolution approach and its advanced compact FPGA implementation architecture. Starting from SISO-FLC for single-branch processing to MIMO-FLC for multiple-branch processing, we will illustrate the key processing units of FLC approach and how to build an efficient FLC processing core on FPGA. A case study will demonstrate how to utilize MIMO-FLC core to construct a compact spatial multiplexing filter bank for Massive MIMO application and baseband equivalent MIMO channel emulator for the forthcoming 5G wireless networks.

5.1 Introduction

Digital linear convolution has its unique territory in many digital signal processing applications, such as image, video signal processing, and communications signal processing. Particularly, coefficients adaptive digital linear convolution is one of the very promising 5G technique candidates for creating spatial multiplexing filter banks in the software defined large scale antenna system (LSAS) or the so-called massive multiple-input multiple output (Massive MIMO) system [1].

The basic concept of LSAS is quite straightforward, namely, we can utilize many radiating antenna elements in conjunction with space division multiplexing (or simply called spatial multiplexing, or precoding) technique to provide wireless access/backhaul/fronthaul services to multiple legacy single-antenna or emerged multiple antenna wireless network equipments, such as multi-functional small cells. Generally, precoding function is achieved/implemented by continuously convoluting data streams with certain coefficient sets, which can be derived in two ways: (1) frequency independent manner, which applies specific space multiplexed codes

L. Guan, *FPGA-based Digital Convolution for Wireless Applications*,
Springer Series in Wireless Technology, DOI 10.1007/978-3-319-52000-1_5

(coefficients) from a pre-defined codebook [2]; (2) frequency dependent manner, which takes the wireless channel response into consideration when doing spatial multiplexing [3]. In the latter case, the wider the bandwidth of signal occupies, the larger the number of coefficients is required. From the concept, it is not difficult to tell that the frequency independent precoding approach is relative simple due to the low-complexity pre-defined coefficients architecture, while the frequency dependent precoding approach faces complexity issue that raised by its nature, i.e., frequency bandwidth related approach.

Previously, in Chaps. 3 and 4, we discussed the digital linear convolution and digital nonlinear convolution. When the number of convolution coefficients goes up, the time-domain convolution implementation architecture suffers from dramatically increased complexity issue, which may slow down the penetration of the new technologies or even block the way to effectively utilize the system level innovation or just simply reduce the attractiveness of the new architectural-level ideas.

In this chapter, we will firstly introduce the so-called Fast Linear Convolution (FLC) approach, and then explain the processing procedures in terms of both overlap save (OLS) manner and overlap add (OLA) manner. Secondly, we will explain the frame work of how to move from single-input single-output FLC (SISO-FLC) to MIMO-FLC. Then we will illustrate how to create a compact MIMO-FLC IP-core on FPGAs. Lastly, we will discuss the extended MIMO-FLC implementation architecture and its appealing applications in wireless communications: (1) the spatial multiplexing filter bank for Massive MIMO system and (2) the baseband equivalent MIMO Channel Emulator.

5.2 SISO-Fast Linear Convolution

The philosophy behind using FLC to replace LC is not complicated, i.e., if we cannot perform a linear operation efficiently in one domain (e.g., time domain), we may think of translating that processing in another transformed domain (e.g., frequency domain) with potential low complexity transformed operation in that domain. Specifically, since the time domain convolution operation is "equivalent" to the multiplication operation in the frequency domain, we can actually carry out frequency domain multiplication instead of doing time domain convolution for some applications, and by doing this we may save quite a bit processing resources. In order to distinguish the latter MIMO-FLC approach, here we will use SISO-FLC to represent the FLC processing for one data stream with corresponding one convolution coefficient set.

The basic approach of SISO-FLC follows a simple route, i.e., we split the continuous data into smaller segments (in the time domain), and we convert both segmental data and coefficients into frequency domain, and we perform the multiplication in the frequency domain and convert the segmental processed data back to time domain, and then re-assemble the processed segmental data to form the final convoluted signal by either discarding redundant samples (OLS manner) or

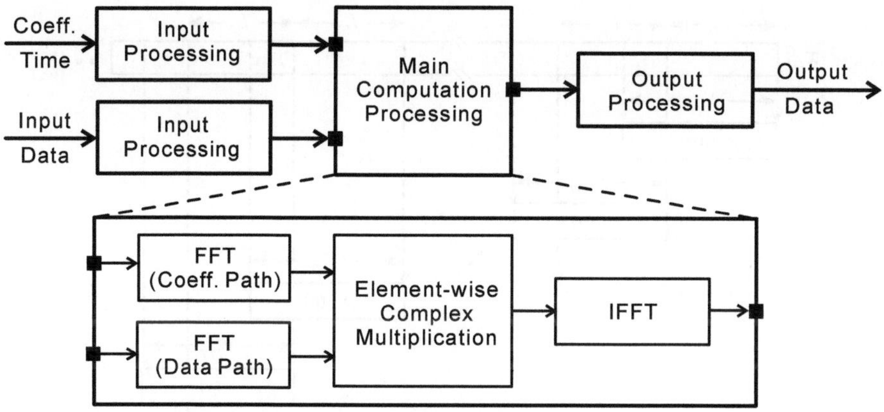

Fig. 5.1 General processing architecture of SISO-FLC

aggregating partial of neighbour segments, (OLA manner). So now, let's have a closer look at what do we need for doing FLC processing. Figure 5.1 illustrates the general processing architecture of SISO-FLC, including input processing units for coefficient path and data path respectively, main computation processing unit, and output processing unit.

5.2.1 Overlap Save Approach

Overlap Save (OLS)-based FLC approach achieves linear convolution function by firstly splitting the incoming data sequence $x_{IN}(n)$ into overlapped sub-sequences $x_i(n)$, each of which is N_{FFT} in length, as represented by Eq. 5.1. Each segment contains L_D effective samples and can be obtained by $L_D = N_{FFT} - N_{Coeff}$, where N_{Coeff} represents the number of convolution coefficients. This input processing for data path is described in Fig. 5.2.

$$x_i(n) = x_{IN}(n + iL_D), n = 0, \ldots, N_{FFT} - 1 \tag{5.1}$$

Secondly, we need to convert time domain segmental data $x_i(n)$ and time domain response, i.e., convolution coefficients $h(n)$ to the frequency domain counterparts, utilizing two FFTs respectively. Next we apply (element-wise multiply) the frequency domain convolution coefficients to the frequency domain data, and then we translate the processed (convoluted) signal from frequency domain to the time domain by doing IFFT function. This processing is described by Eq. 5.2,

$$y_i(n) = \mathbf{IFFT}\{\mathbf{FFT}[x_i(n)] \odot \mathbf{FFT}[h_{zp}(n)]\} \tag{5.2}$$

$h_{zp}(n)$ is zero-padded version of response $h(n)$, namely, $h_{zp}(n) = [h(n), 0, \ldots, 0]$, where the number of zeros padded is equal to the difference between N_{FFT} and the

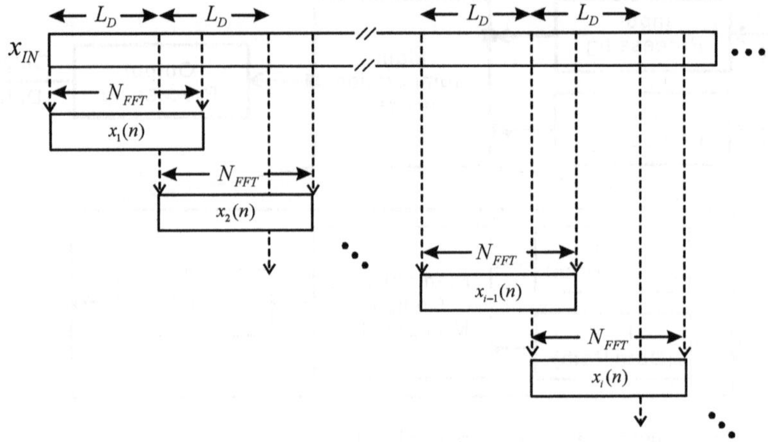

Fig. 5.2 FLC input processing illustration when using OLS approach

actual number of response samples, i.e., L_D. The symbol $\odot$ denotes element-wise complex multiplication operation: given two complex data sequences of length N, e.g., $a_i(n) = \left[a_i^1, a_i^2, \ldots, a_i^N\right]$ and $b_k(n) = \left[b_k^1, b_k^2, \ldots, b_k^N\right]$, the element-wise complex multiplication operator is performed as Eq. 5.3,

$$a_i(n) \odot b_k(n) = \left[a_i^1 b_k^1, a_i^2 b_k^2, \ldots, a_i^N b_k^N\right], a_i^n \quad \text{and} \quad b_k^n \in \mathbf{C} \tag{5.3}$$

And lastly, we re-construct the processed data according to the original timing relationship and form the output data sequence $y_{OUT}(n)$ by discarding the first overlapped $N_{FFT} - L_D$ data samples as Eq. 5.4, (Fig. 5.3)

$$y_{OUT}(n + iL_D) = y_i(n + N_{FFT} - L_D),\ n = 0,\ \ldots,\ L_D - 1 \tag{5.4}$$

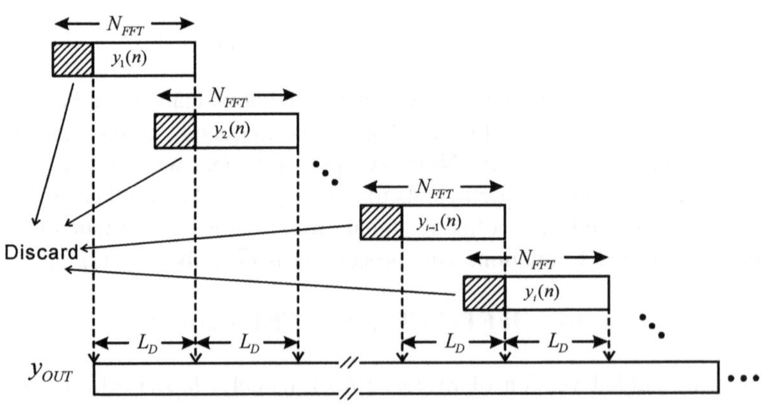

Fig. 5.3 FLC output processing illustration when using OLS approach

5.2.2 Overlap Add Approach

Overlap Add (OLA)—based FLC approach achieves linear convolution function by firstly splitting the incoming data sequence $x_{IN}(n)$ into non-overlapped sub-sequences $x_i(n)$, each of which is N_{FFT} in length, containing L_D effective samples and $N_{FFT} - L_D$ zeros, as represented by Eq. 5.5. The corresponding processing is illustrated in Fig. 5.4.

$$x_i(n) = \begin{cases} x_{IN}(n + iL_D) & n = 0, \ldots, L_D - 1 \\ 0 & , n = L_D, \ldots, N_{FFT} - 1 \end{cases} \tag{5.5}$$

Secondly, we will to do the same computation processing as OLS approach does, namely, we convert time domain segmental data $x_i(n)$ and time domain response, i.e., convolution coefficients $h(n)$ to the frequency domain counterparts, utilizing two FFTs respectively. Next we apply the frequency domain convolution coefficients to the frequency domain data, and then we translate the processed (convoluted) signal from frequency domain to the time domain by doing IFFT function. This processing has been described by Eq. 5.2.

And lastly, we re-construct the processed data according to the original timing relationship and form the output data sequence $y_{OUT}(n)$ by aggregating the first overlapped $N_{FFT} - L_D$ data samples as Eq. 5.6. And Fig. 5.5 illustrates the processing for the output re-construction.

$$y_{OUT}(n + iL_D) = \begin{cases} y_i(n) + y_{i-1}(n + L_D), & n = 0, \ldots, N_{FFT} - L_D - 1 \\ y_i(n), & n = N_{FFT} - L_D, \ldots, N_{FFT} - 1 \end{cases} \tag{5.6}$$

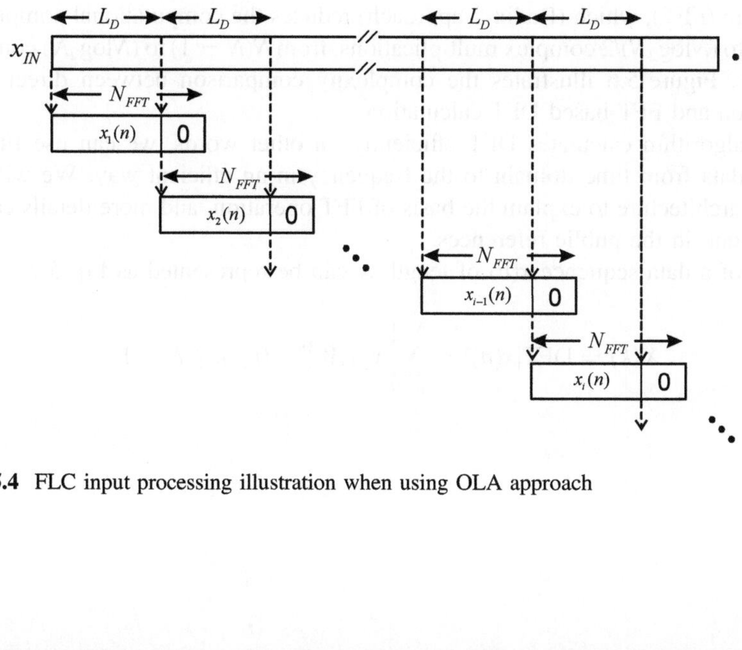

Fig. 5.4 FLC input processing illustration when using OLA approach

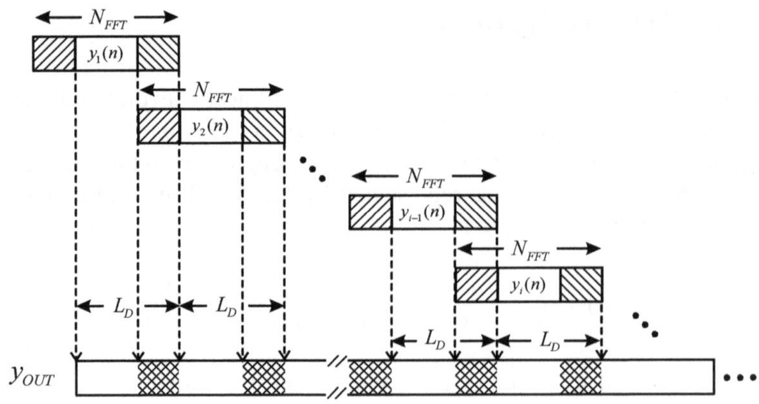

Fig. 5.5 FLC output processing illustration when using OLS approach

5.2.3 FFT Basis

From the description of SISO-FLC, we can see, the key the processing is the transformation function, which is used to transform the data between time domain and frequency domain.

Discrete Fourier Transform (DFT) is a very powerful signal analysis approach, which transforms the signal from time domain to the so-called frequency domain. However, the computational complexity of DFT operation will increase dramatically (following the square law) when the number of DFT processing samples are increasing. For instance, N-point DFT needs N^2 complex multiplication operations and $N(N-1)$ complex addition operations. Its inherent complexity issue actually limited its applications in the real world. In 1965, Cooley and Turkey [4] proposed a very efficient DFT calculation algorithm, namely the so-called Fast Fourier Transform (FFT), which (Radix-2 approach) reduces the computational complexity from N^2 to $(N\log_2 N)/2$ complex multiplications, from $N(N-1)$ to $(N\log_2 N)$ complex additions. Figure 5.6 illustrates the complexity comparison between direct DFT calculation and FFT-based DFT calculation.

FFT algorithm calculates DFT efficiently, in other words, we can use FFT to convert data from time domain to the frequency in an efficient way. We will use Radix-2 architecture to explain the basis of FFT operation, and more details can be easily found in the public references.

DFT of a data sequence $x(n)$ of length N can be represented as Eq. 5.7,

$$X(k) = \mathrm{DFT}[x(n)] = \sum_{n=0}^{N-1} x(n) W_N^{kn} \quad 0 \le k \le N-1 \tag{5.7}$$

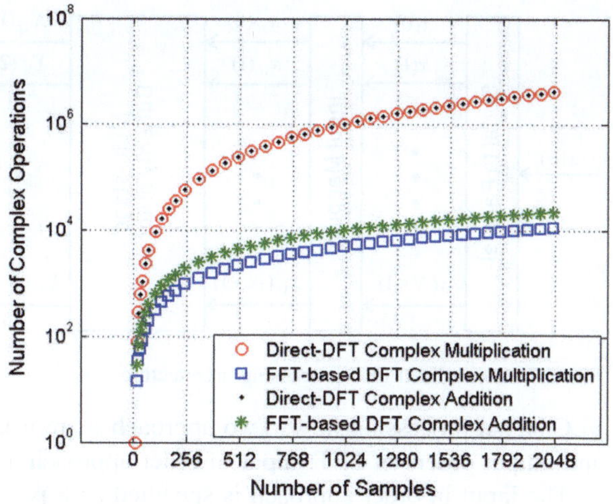

Fig. 5.6 Computational complexity comparison between direct DFT and FFT-based DFT

And its corresponding inverse transform IDFT is defined as Eq. 5.8,

$$x(n) = \mathrm{IDFT}[X(k)] = \frac{1}{N}\sum_{k=0}^{N-1} X(k)W_N^{-kn} \quad 0 \leq k \leq N-1 \tag{5.8}$$

where $W_N^{kn} = \mathrm{e}^{-j2\pi kn/N}$.

If we tear down the rotation factor W_N^k it is not difficult to realize that this factor is periodic and symmetrical as shown in Eqs. 5.9 and 5.10, respectively.

$$W_N^{k+N} = \mathrm{e}^{-j2\pi(k+N)/N} = \mathrm{e}^{-2\pi k/N} \cdot \mathrm{e}^{-j2\pi} = \mathrm{e}^{-2\pi k/N} = W_N^k \tag{5.9}$$

$$W_N^{k+N/2} = \mathrm{e}^{-j2\pi(k+N/2)/N} = \mathrm{e}^{-2\pi k/N} \cdot \mathrm{e}^{-j\pi} = -\mathrm{e}^{-2\pi k/N} = -W_N^k \tag{5.10}$$

Euler's formula was used in both Eqs. 5.9 and 5.10 as below,

$$\mathrm{e}^{-j2\pi} = \cos(-2\pi) + j\ \sin(-2\pi) = 1 \tag{5.11}$$

$$\mathrm{e}^{-j\pi} = \cos(-\pi) + j\sin(-\pi) = -1 \tag{5.12}$$

Radix-2 FFT algorithm actually takes the advantages periodic and symmetrical property of W_N^k, i.e., Eqs. 5.9 and 5.10, when performing DFT calculation, so that some complex multiplication operations can be shared. As shown in Fig. 5.7, a fully functional FFT calculation engine includes the following processing units: serial to parallel unit, input mapping unit, butterfly arithmetic unit, output mapping unit, and parallel to serial unit.

For the purpose of building an FLC processing core, we are interested in one of the FFT properties, i.e., the input and output orders, which have the impacts on the

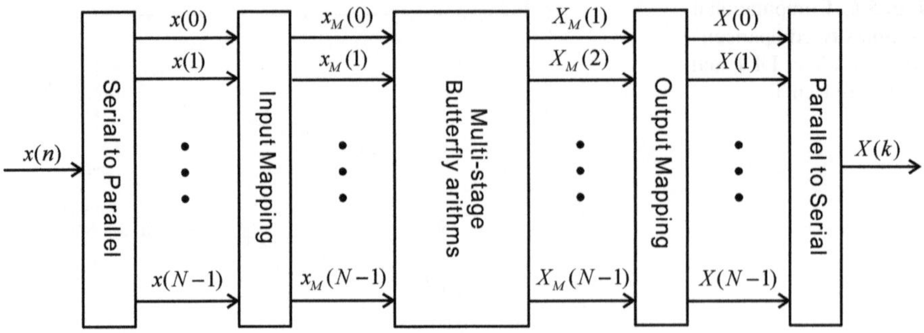

Fig. 5.7 Generalized FFT processing architecture

FLC signal processing flow. Two approaches are usually used regarding the input and output orders of FFT: input in order approach and output in order approach.

The input in order approach is specified as Eqs. 5.13 and 5.14,

$$x\left[(b_l b_{l-1} \ldots b_1 b_0)_2\right] \xrightarrow{\text{Input Mapping}} x_M\left[(b_l b_{l-1} \ldots b_1 b_0)_2\right] \tag{5.13}$$

$$X_M\left[(b_l b_{l-1} \ldots b_1 b_0)_2\right] \xrightarrow{\text{Onput Mapping}} X\left[(b_0 b_1 \ldots b_{l-1} b_l)_2\right] \tag{5.14}$$

where b_0, b_1, …, b_l represents binary value (1 or 0), and operation $(*)_2$ stands for binary representation, e.g., $(101)_2$ is equal to 5 in decimal representation. Table 5.1 illustrates input and output mapping relationship for the input in order approach, using 16-point FFT as an example.

And output in order approach can be described by Eqs. 5.15 and 5.16. Table 5.2 illustrates input and output mapping relationship for the output in order approach, using 16-point FFT as an example.

$$x\left[(b_l b_{l-1} \ldots b_1 b_0)_2\right] \xrightarrow{\text{Input Mapping}} x_M\left[(b_0 b_1 \ldots b_{l-1} b_l)_2\right] \tag{5.15}$$

$$X_M\left[(b_l b_{l-1} \ldots b_1 b_0)_2\right] \xrightarrow{\text{Onput Mapping}} X\left[(b_l b_{l-1} \ldots b_1 b_0)_2\right] \tag{5.16}$$

There are quite a few good explanations and references of radix-2 FFT algorithm using decimation in time (DIT) and decimation in frequency (DIF) approaches, respectively [5]. Figure 5.8 illustrates the computation difference between DIT-FFT and DIF-FFT approaches.

Both DIF-FFT and DIT-FFT can be arranged as input in order or output in order with the same computational complexity. Since the detailed FFT algorithm is not the focus of this book, here, we directly give two examples of 8-point FFT signal flow charts to provide you a flavour of how FFT unit is running. Figure 5.9

Table 5.1 Input and output mapping example (input in order)

Binary				Index		Input In Order			
b_3	b_2	b_1	b_0	$\rightarrow$ MSB b_3	$\leftarrow$ MSB b_0	Input Mapping		Output Mapping	
0	0	0	0	0	0	$x(0)$	$x_M(0)$	$X_M(0)$	$X(0)$
0	0	0	1	1	8	$x(1)$	$x_M(1)$	$X_M(1)$	$X(8)$
0	0	1	0	2	4	$x(2)$	$x_M(2)$	$X_M(2)$	$X(4)$
0	0	1	1	3	12	$x(3)$	$x_M(3)$	$X_M(3)$	$X(12)$
0	1	0	0	4	2	$x(4)$	$x_M(4)$	$X_M(4)$	$X(2)$
0	1	0	1	5	10	$x(5)$	$x_M(5)$	$X_M(5)$	$X(10)$
0	1	1	0	6	6	$x(6)$	$x_M(6)$	$X_M(6)$	$X(6)$
0	1	1	1	7	14	$x(7)$	$x_M(7)$	$X_M(7)$	$X(14)$
1	0	0	0	8	1	$x(8)$	$x_M(8)$	$X_M(8)$	$X(1)$
1	0	0	1	9	9	$x(9)$	$x_M(9)$	$X_M(9)$	$X(9)$
1	0	1	0	10	5	$x(10)$	$x_M(10)$	$X_M(10)$	$X(5)$
1	0	1	1	11	13	$x(11)$	$x_M(11)$	$X_M(11)$	$X(13)$
1	1	0	0	12	3	$x(12)$	$x_M(12)$	$X_M(12)$	$X(3)$
1	1	0	1	13	11	$x(13)$	$x_M(13)$	$X_M(13)$	$X(11)$
1	1	1	0	14	7	$x(14)$	$x_M(14)$	$X_M(14)$	$X(7)$
1	1	1	1	15	15	$x(15)$	$x_M(15)$	$X_M(15)$	$X(15)$

illustrates DIT-FFT with output in order approach and Fig. 5.10 illustrates DIF-FFT with input in order approach.

Due to the effectiveness of Cooley-Tukey FFT algorithm for calculating DFT, multiple vendors have designed and created hardware-based FFT IP-cores, which can be easily used in different signal processors. For example, Xilinx FFT IP-core [6] implements the Cooley-Tukey FFT algorithm for both burst radix-2 and radix-4 I/O architectures with DIT method and pipelined streaming I/O with DIF method. More information is available from Xilinx official website.

5.2.4 SISO-FLC Complexity

In this section, we will evaluate the SISO-FLC algorithm regarding computational complexity. According to the FLC processing diagram Fig. 5.1, complex multiplication operations are required only in the computation processing unit, which including FFT in the data path, FFT in the coefficient path, element-wise complex multiplication (CM), and IFFT in the data path. Here, the length of time domain convolution coefficients and FFT are denoted by N_{Coeff} and N_{FFT} ($N_{FFT} > N_{Coeff}$),

Table 5.2 Input and output mapping example (output in order)

Binary				Index		Output In Order			
b_3	b_2	b_1	b_0	→ MSB b_3	← MSB b_0	Input Mapping		Output Mapping	
0	0	0	0	0	0	$x(0)$	$x_M(0)$	$X_M(0)$	$X(0)$
0	0	0	1	1	8	$x(1)$	$x_M(8)$	$X_M(1)$	$X(1)$
0	0	1	0	2	4	$x(2)$	$x_M(4)$	$X_M(2)$	$X(2)$
0	0	1	1	3	12	$x(3)$	$x_M(12)$	$X_M(3)$	$X(3)$
0	1	0	0	4	2	$x(4)$	$x_M(2)$	$X_M(4)$	$X(4)$
0	1	0	1	5	10	$x(5)$	$x_M(10)$	$X_M(5)$	$X(5)$
0	1	1	0	6	6	$x(6)$	$x_M(6)$	$X_M(6)$	$X(6)$
0	1	1	1	7	14	$x(7)$	$x_M(14)$	$X_M(7)$	$X(7)$
1	0	0	0	8	1	$x(8)$	$x_M(1)$	$X_M(8)$	$X(8)$
1	0	0	1	9	9	$x(9)$	$x_M(9)$	$X_M(9)$	$X(9)$
1	0	1	0	10	5	$x(10)$	$x_M(5)$	$X_M(10)$	$X(10)$
1	0	1	1	11	13	$x(11)$	$x_M(13)$	$X_M(11)$	$X(11)$
1	1	0	0	12	3	$x(12)$	$x_M(3)$	$X_M(12)$	$X(12)$
1	1	0	1	13	11	$x(13)$	$x_M(11)$	$X_M(13)$	$X(13)$
1	1	1	0	14	7	$x(14)$	$x_M(7)$	$X_M(14)$	$X(14)$
1	1	1	1	15	15	$x(15)$	$x_M(15)$	$X_M(15)$	$X(15)$

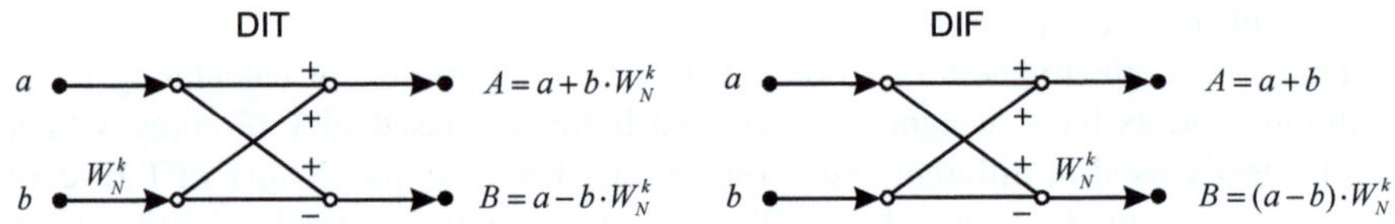

Fig. 5.8 Comparison between DIT and DFT calculation approach

respectively. Table 5.3 summaries the complexity of each sub-unit regarding complex multiplication (CM) and complex addition (CA).

For simplicity, we will focus the number of CMs only (since comparing to CA, CM is more complicate and occupying relative larger processing resources on the silicon). For data path, the complexity $O(\text{SISO - FLC})$ that is required to calculate one output sample by SISO-FLC algorithm is given by Eq. 5.17,

$$O(\text{SISO - FLC}) = \frac{\frac{N_{FFT}}{2}\log_2(N_{FFT})\cdot 2 + N_{FFT}}{N_{FFT} - N_{Coeff}} \tag{5.17}$$

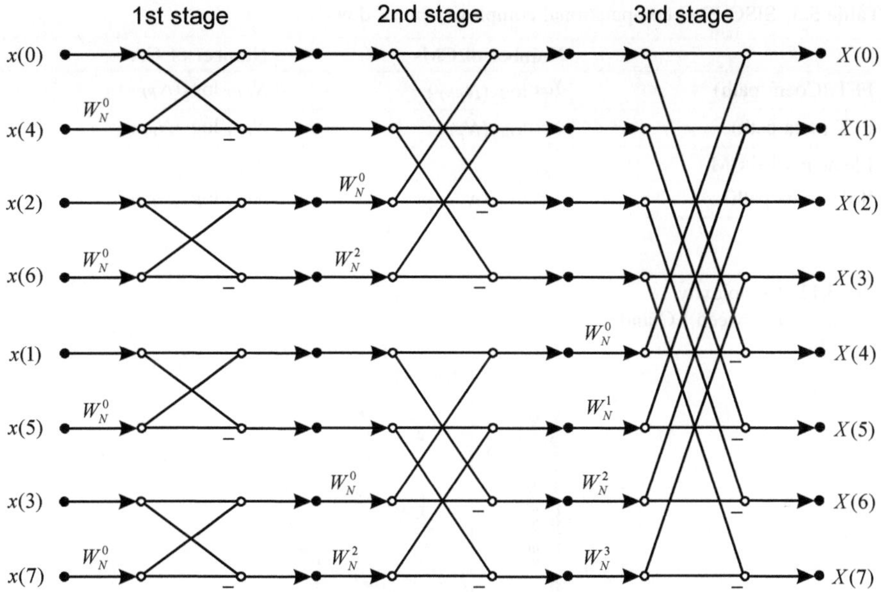

Fig. 5.9 8-point FFT calculation flow chart in a DIT manner

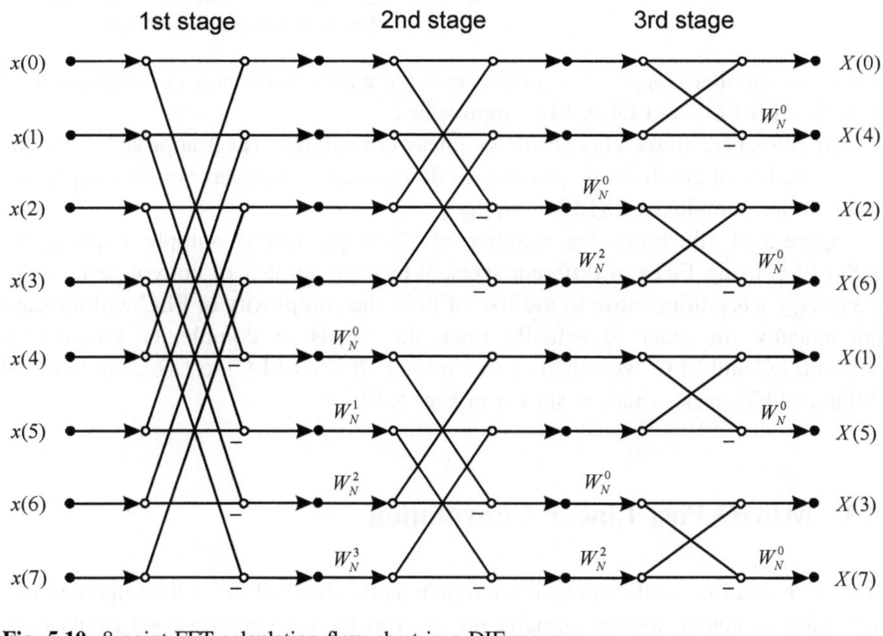

Fig. 5.10 8-point FFT calculation flow chart in a DIF manner

Table 5.3 SISO-FLC computational complexity break-down

	Number of CMs	Number of CAs
FFT (Coeff. path)	$\frac{N_{FFT}}{2}\log_2(N_{FFT})$	$N_{FFT}\log_2(N_{FFT})$
FFT (data path)	$\frac{N_{FFT}}{2}\log_2(N_{FFT})$	$N_{FFT}\log_2(N_{FFT})$
Element-wise CM	N_{FFT}	0
IFFT (data path)	$\frac{N_{FFT}}{2}\log_2(N_{FFT})$	$N_{FFT}\log_2(N_{FFT})$

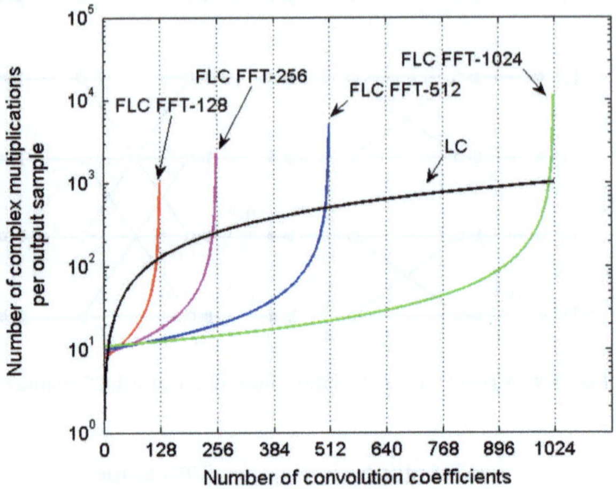

Fig. 5.11 Complexity comparison between LC and SISO-FLC

where the number $N_{FFT} - N_{Coeff}$ represents the effective samples per segment used in both OLS-FLC and OLA-FLC approaches.

For reference, if we choose direct linear convolution (LC) approach with the same number of coefficients, in order to derive one convoluted output sample, we will require complexity $O(\text{LC}) = N_{Coeff}$.

Figure 5.11 illustrates the number of CMs per output sample required by SISO-FLC using FFTs of different sizes. When the number of convolution coefficients N_{Coeff} is getting close to the size of FFT, the complexity of FLC will increase dramatically. In order to actually track the trends of complexity relationship between LC and FLC, we calculate the ratio of O(SISO-FLC) to O(LC) in terms of different FFT sizes, which is shown in Fig. 5.12.

5.3 MIMO-Fast Linear Convolution

If the convolution coefficients are not dynamically changed (or will be updated in a very slow manner), we can actually pre-convert the convolution coefficients from time domain to frequency domain and store the frequency domain coefficients in some types of memories. Thus, SISO-FLC approach that deals with single-branch convolution will only need real-time FFT and IFFT in the continuously processed

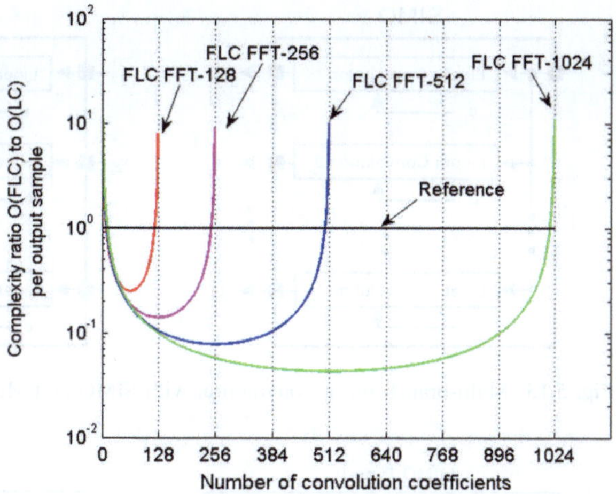

Fig. 5.12 Complexity ratio comparison among SISO-FLC with different FFT sizes

data path (no FFT is required in the coefficient path due to the pre-conversion of the coefficients). In this way, we can save valuable real-time processing resources. However, some applications will need multiple interactive convolutions (like Massive MIMO system, which will be discussed later in this chapter), and the simple duplication of SISO-FLC in parallel strategy doesn't offer an efficient solution, we need to explore some optimal processing architectures for the application that requires multi-branch adaptive long-tap convolutions.

Luckily, one of the most important natures of FPGA is parallel processing, which differentiates FPGA from other processors fundamentally. In other words, multi-branch FLC will fit in well on the FPGA platform naturally. So, let's have a closer look at what do we need to do for efficiently building multiple interactive FLCs on FPGAs.

5.3.1 From SISO to MIMO

As shown in Fig. 5.13, single-input multiple-output (SIMO) convolution system can be achieved by placing multiple linear convolutions in parallel with same input and different coefficient sets for different output branches. Similarly, multiple-input single-output (MISO) convolution system is shown in the figure as well, where we use a simple addition to illustrate the interactive function by aggregating multiple convolutions together.

Without loss of generalities, an example of multiple-input multiple-output (MIMO) convolution system is depicted in Fig. 5.14. In Type-I MIMO system, M inputs will be convoluted by $M \times N$ coefficients sets, and aggregated by N parallel additions, forming N interactive convoluted outputs. The parameter $\mathbf{c}^{\mathrm{I}}_{m,n}(1 \le m \le M,\ 1 \le n \le N)$ represents the time domain coefficient set corresponding to the response between mth input and nth output. And in Type-II MIMO

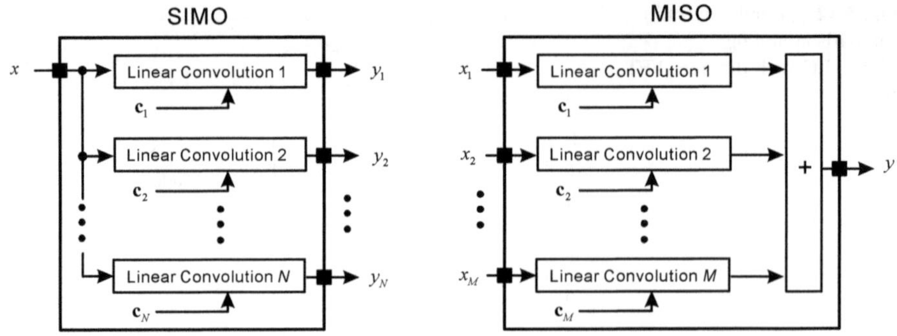

Fig. 5.13 Multi-branch linear convolution with SIMO and MISO architecture

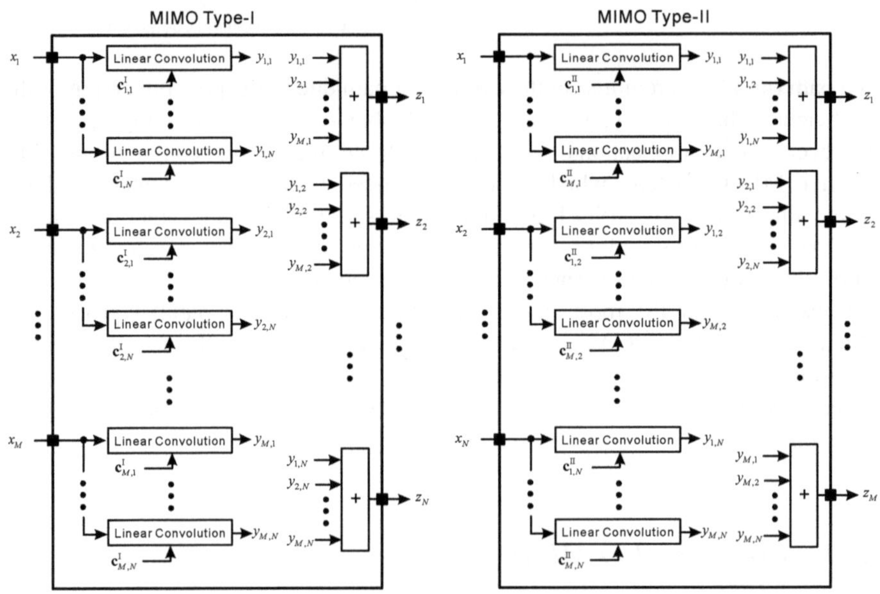

Fig. 5.14 An example of generalized MIMO linear convolution architecture (Type-I: M to N MIMO, Type-II: N to M MIMO)

system, N inputs will be convoluted by $N \times M$ coefficients sets, and aggregated by M parallel additions, forming M interactive convoluted output. The parameter $\mathbf{c}^{\mathrm{II}}_{m,n}(1 \leq m \leq M,\ 1 \leq n \leq N)$ represents the time domain coefficient set corresponding to the response between nth input and mth output. $y_{m,n}$ in Type-I architecture and Type-II architecture are corresponding to the output data of convolution x_m with $\mathbf{c}^{\mathrm{I}}_{m,n}$ and the output data of convolution x_n with $\mathbf{c}^{\mathrm{II}}_{m,n}$, respectively. The difference between Type-I and Type-II is the arrangement of coefficient sets. For instance, let's construct the coefficient sets in the matrix format $[\mathbf{c}_{m,n}]$, Type-I MIMO will use the mth row of the coefficient set matrix to represent the response between the mth input and all outputs, while Type-II MIMO will use the nth column of the coefficient set matrix to represent the response between the nth input and all outputs. Mathematically, the coefficients of Type-I MIMO and Type-II MIMO architecture will follow as $\mathbf{c}^{\mathrm{I}}_{m,n} = \mathbf{c}^{\mathrm{II}}_{n,m}$. The definition of Type-I MIMO and Type-II MIMO will be used to facilitate the MIMO-FLC processing explanation in the following sections. However if we treat Type-I MIMO and Type-II MIMO as black boxes, and we are only interested in the inputs and outputs of the black box, there is no fundamental differences between those two architectures.

For the purpose of efficiently building an interactive MIMO convolution system, we can replace the each linear convolution unit in the Fig. 5.14 by FLC unit straightforwardly. And the complexity of this direct approach will be multiple times of the SISO-FLC complexity as described by Fig. 5.12. However, in order to better utilize the FPGA with its high-speed parallel processing capability, we need some technical tricks to create a more suitable architecture for more efficiently MIMO-FLC processing, which will be described in the coming section.

5.3.2 Unit Decomposition and Sub-function Sharing

Let's recall the general FLC processing architecture in Fig. 5.1, which includes three computation units, i.e., FFT, element-wise complex multiplication, and IFFT. For the sake of simplicity, we assume the time domain convolution coefficients are static, and the corresponding frequency domain coefficients have been pre-calculated in a ready-to-use format. Before "convoluted" by frequency domain coefficients, the data will be transformed to frequency domain by FFT, and same time domain data stream will be corresponding to the same FFT data stream. So in the direct MIMO convolution system architecture, there is certain level redundancy during the processing stages. In other words, some data have been processed in the same way by different parallel processing units. Motivated by this, we can decompose the SISO-FLC unit into multiple sub-functional units, and share some of the sub-functions among multiple parallel processing stages to get rid of redundant processing units. For example, the following arrangement would be some good choices to reduce the complexity of MIMO-FLC system:

- Decomposing FLC processing into 5 processing stages including input buffer stage, FFT stage, element-wise complex multiplication (E-CM) stage, IFFT stage and output buffer stage;
- Utilizing one input buffer unit, together with one followed FFT unit for the same input data stream;
- Aggregating multiple convoluted data streams in the frequency, i.e., moving aggregation unit from after output buffer stage to before IFFT stage;
- Utilizing one IFFT unit, together with one followed output buffer unit for the same aggregated output data stream.

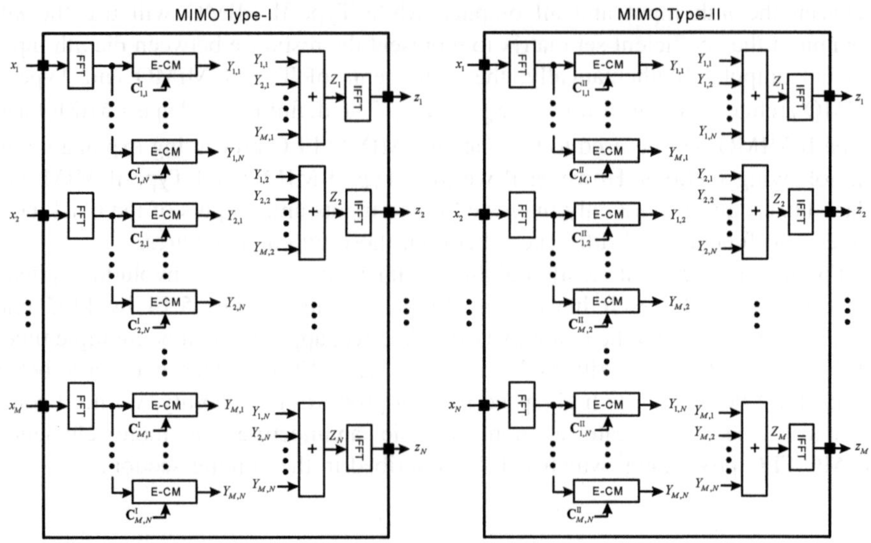

Fig. 5.15 Improved MIMO-FLC processing architecture using sub-function sharing strategy

Based on the unit decomposition and sub-function sharing strategy described above, we outline a more effective architecture to process MIMO-FLC. Figure 5.15 shows the diagram of the improved MIMO-FLC architecture, where $\mathbf{C}^{\mathrm{I}}_{m,n} = \mathbf{FFT}\left[\mathbf{c}^{\mathrm{I}}_{m,n}, 0, \ldots, 0\right], \quad (1 \le m \le M, 1 \le n \le N)$ represents the frequency domain coefficient set of Type-I MIMO. And the Type-II MIMO frequency domain coefficient set $\mathbf{C}^{\mathrm{II}}_{m,n}$ is generated in the similar approach by FFT.

The optimized arrangement described above architecturally reduces the complexity of MIMO-FLC, Table 5.4 summaries the complexity comparison between direct MIMO-FLC architecture and improved MIMO-FLC architecture. Here, in Type-I architecture M inputs will pass the MIMO-FLC unit generating N interactive convoluted outputs, and in Type II architecture, N inputs will pass the MIMO-FLC unit generating M interactive convoluted outputs.

Table 5.4 Complexity comparison between direct MIMO-FLC architecture and improved MIMO-FLC architecture (both Type-I and Type-II)

	Direct MIMO-FLC architecture	Improved MIMO-FLC architecture
Number of FFT unit	$M \times N$	M (type-I) or N (type-II)
Number of IFFT unit	$M \times N$	N (type-I) or M (type-II)
Number of E- CM unit	$M \times N$	$M \times N$
Total number of CMs	Equation 5.18	Equation 5.19

If we take the complexity of FFT/IFFT and E-CM into consideration, the actual number of complex multiplication operations for direct MIMO-FLC architecture $\left(O_{\text{MIMO-FLC}}^{\text{Direct}}\right)$ and improved MIMO-FLC architecture $\left(O_{\text{MIMO-FLC}}^{\text{Improved}}\right)$ are given by Eqs. 5.18 and 5.19, respectively.

$$\begin{aligned} O_{\text{MIMO-FLC}}^{\text{Direct}} &= \frac{N_{FFT}}{2}\log_2(N_{FFT})\cdot 2\cdot M\cdot N + N_{FFT}\cdot M\cdot N \\ &= N_{FFT}\cdot[\log_2(N_{FFT})\cdot M\cdot N + M\cdot N] \end{aligned} \tag{5.18}$$

$$\begin{aligned} O_{\text{MIMO-FLC}}^{\text{Improved}} &= \frac{N_{FFT}}{2}\log_2(N_{FFT})\cdot (M+N) + N_{FFT}\cdot M\cdot N \\ &= N_{FFT}\cdot\left[\log_2(N_{FFT})\cdot\frac{M+N}{2} + M\cdot N\right] \end{aligned} \tag{5.19}$$

Comparing Eqs. 5.18 and 5.19, only when $M = 1$ and $N = 1$, improved MIMO-FLC has the same complexity as direct MIMO-FLC, otherwise, improved MIMO-FLC architecture will always have lower complexity given the meaningful values for M and N.

5.3.3 *Buffered Segment-Level Interleaving and De-interleaving*

In a modern FPGA, the reliable processing rate is usually higher than the required data sampling rate for wireless applications. For example, the baseband data rate of one 20 MHz LTE component carrier in the 4G wireless communication system is 30.72 Mega sample per second (MSPS), while the state-of-the-art FPGA can reliably operate at couple of hundred MHz, which is much higher than 30.72 MHz. As introduced before, the gap between required data rate and available processing rate provides an opportunity to oversample the signal and utilize the so-called time division multiplexing (TDM) technique to reduce the actual number of physical resources. This TDM-based resource sharing concept can be extended to data segment-level, i.e., if we are using the processing rate that is higher than data rate, the original continuous slow-rate data can be logically split into high-speed bursty data segments, and between two bursty data segments there are some processing gaps which can be efficiently fitted by another (or multiple) bursty data segments from different data branches at the same processing stage.

The processing described above is a segment-level data interleaving approach, which encapsulates multiple slow-rate continuous data streams into a high-speed bursty data stream. For example, as shown in Fig. 5.16, four continuous slow-rate data streams are buffered into bursty high-speed data. During the period of one slow-rate data segment (SS), four high-speed data segments (HS) can appear without overlapping. In other words, we can formulate a busy data channel with

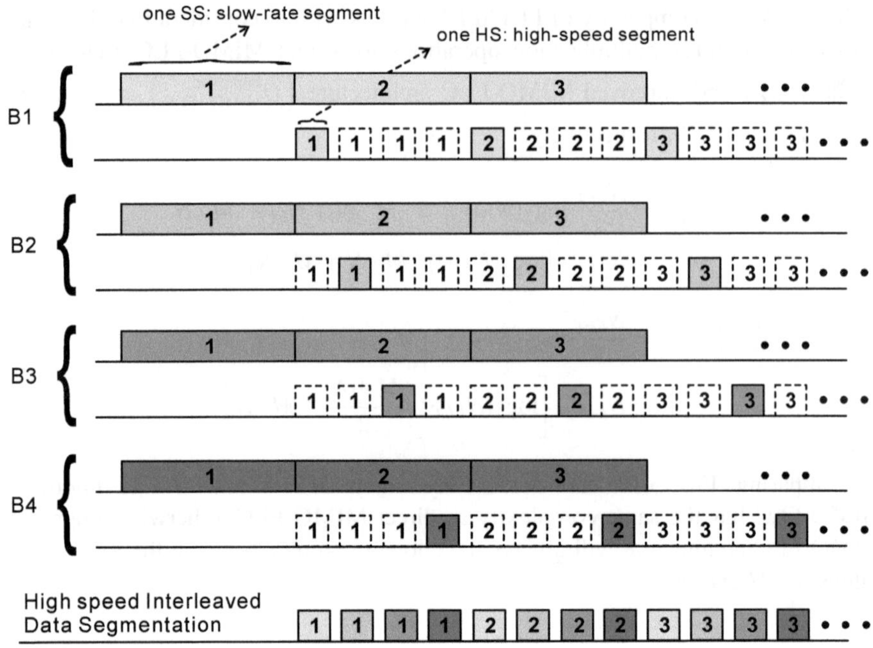

Fig. 5.16 Example of data segment level interleaving processing

interleaved data segments from four branches without losing information or introducing redundant information, shown at the bottom of Fig. 5.16.

This segment-level interleaving technique is quite suitable for efficiently implement MIMO-FLC on the FPGA platform. For instance, by utilizing this approach, multiple inputs can be buffered and interleaved into one high-speed bursty data channel. And only one FFT unit is required for converting this one data channel from time domain to frequency domain. This optimized arrangement is equivalent to utilize one FFT to process multiple data streams. The maximum theoretical interleaving factor is denoted as $N_{Interleave}$ which is determined by Eq. 5.20.

$$N_{Interleave} = floor\left(\frac{N_{FFT} - N_{Coeff}}{N_{FFT}} \cdot \frac{CLK_{FPGA}}{CLK_{DATA}}\right) \tag{5.20}$$

The function *floor* (*) rounds a number to the next smaller integer. And the parameters CLK_{DATA} and CLK_{FPGA} represent the data sampling rate and FPGA processing rate, respectively. The expression ($N_{FFT} - N_{Coeff}$) represents the effective number of samples per data segment in the FLC sense.

When we retrieve the data, two buffered de-interleaving architectures can be used for splitting the FFT processed data to 2-dimensional (2D) signal. The buffered de-interleaving module (for Type-I MIMO) creates several data channels which

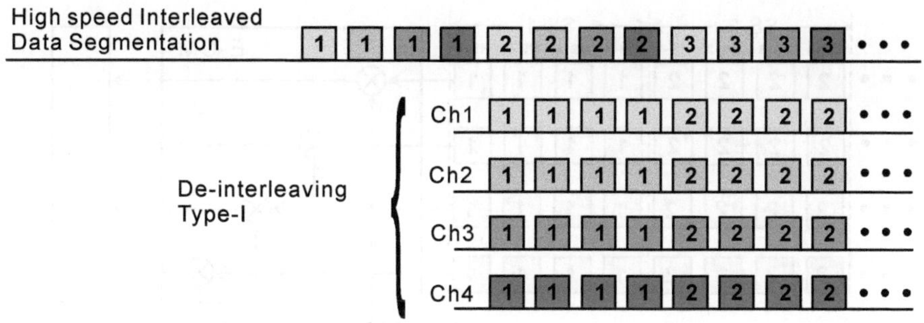

Fig. 5.17 Example of Type-I buffered de-interleaving processing

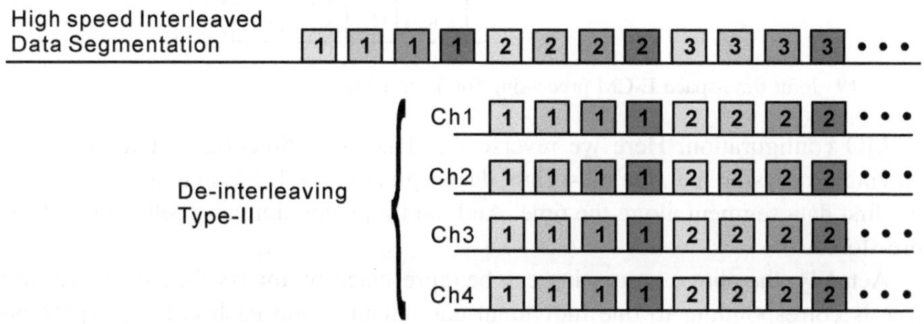

Fig. 5.18 Example of Type-II buffered de-interleaving processing

contains different segments (at the same channel within a period of slow-rate segment, the high-speed data segments are the same), while the buffered de-interleaving module (for Type-II MIMO) generates several data channels which are the same along the time (at the same time slot, different data channels have the same data segment). Figures 5.17 and 5.18 illustrate the buffered de-interleaving processing in Type-I and Type-II respectively (interleave factor $N_{Interleave} = 4$).

5.3.4 Time-Space 2D Signal Processing

Following last sub-section, one high-speed FFT unit will process the high-speed interleaved data segments, i.e., converting time domain data segments to the corresponding frequency domain data segments. Before feeding to the E-CM unit, depending on the MIMO type (Type-I or Type-II), two data segment arrangement options can be adopted for facilitating E-CM operations. Both of those two options will create some joint time-space (2-D) data segments with certain level redundant information, which will be used for creating *M* to *N* (Type-I) or *N* to *M* (Type-II) mapping. Figure 5.19 illustrates the joint time-space data arrangement for Type-I

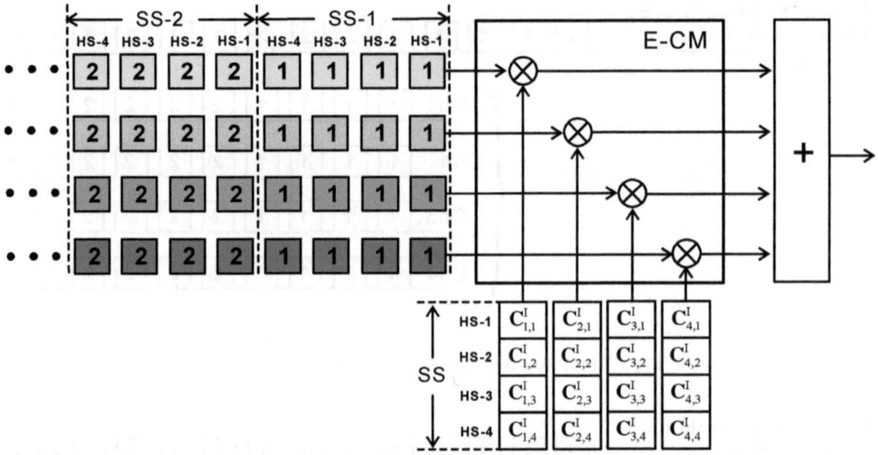

Fig. 5.19 Joint time-space E-CM processing for Type-I MIMO

MIMO configuration. Here we reverse the data flow direction of buffered interleaving processing, i.e., the most right data segment (HS-1, SS-1) in the Fig. 5.19 is the first data segment along the time. And the frequency domain coefficient sets are pre-stored.

Actually, the above processing can be represented by matrix format, where each row is corresponding to one individual data channel and each column represents corresponding high-speed data segment (HS). One SS period processing shown in the Fig. 5.19 can be then represented by Eq. 5.21, using 2D E-CM and column aggregation.

$$
\begin{aligned}
&\sum^{\text{column}} \begin{bmatrix} X_{1,1}\odot \mathbf{C}^{\mathrm{I}}_{1,1} & X_{1,1}\odot \mathbf{C}^{\mathrm{I}}_{1,2} & X_{1,1}\odot \mathbf{C}^{\mathrm{I}}_{1,3} & X_{1,1}\odot \mathbf{C}^{\mathrm{I}}_{1,4} \\ X_{2,1}\odot \mathbf{C}^{\mathrm{I}}_{2,1} & X_{2,1}\odot \mathbf{C}^{\mathrm{I}}_{2,2} & X_{2,1}\odot \mathbf{C}^{\mathrm{I}}_{2,3} & X_{2,1}\odot \mathbf{C}^{\mathrm{I}}_{2,4} \\ X_{3,1}\odot \mathbf{C}^{\mathrm{I}}_{3,1} & X_{3,1}\odot \mathbf{C}^{\mathrm{I}}_{3,2} & X_{3,1}\odot \mathbf{C}^{\mathrm{I}}_{3,3} & X_{3,1}\odot \mathbf{C}^{\mathrm{I}}_{3,4} \\ X_{4,1}\odot \mathbf{C}^{\mathrm{I}}_{4,1} & X_{4,1}\odot \mathbf{C}^{\mathrm{I}}_{4,2} & X_{4,1}\odot \mathbf{C}^{\mathrm{I}}_{4,3} & X_{4,1}\odot \mathbf{C}^{\mathrm{I}}_{4,4} \end{bmatrix} \\
&= \left[\sum_{m=1}^{4}\left(X_{m,1}\odot \mathbf{C}^{\mathrm{I}}_{m,1}\right) \quad \sum_{m=1}^{4}\left(X_{m,1}\odot \mathbf{C}^{\mathrm{I}}_{m,2}\right) \quad \sum_{m=1}^{4}\left(X_{m,1}\odot \mathbf{C}^{\mathrm{I}}_{m,3}\right) \quad \sum_{m=1}^{4}\left(X_{m,1}\odot \mathbf{C}^{\mathrm{I}}_{m,4}\right) \right] \\
&= \left[\sum_{m=1}^{4}(Y_{m,1}) \quad \sum_{m=1}^{4}(Y_{m,2}) \quad \sum_{m=1}^{4}(Y_{m,3}) \quad \sum_{m=1}^{4}(Y_{m,4}) \right] \\
&= [Z_{1,1} \quad Z_{2,1} \quad Z_{3,1} \quad Z_{4,1}]
\end{aligned}
\tag{5.21}
$$

where $X_{m,i}$ represents the ith frequency domain data segment at mth branch, $Y_{m,n}$ is corresponding to the frequency domain data of convolution X_m with $\mathbf{C}^{\mathrm{I}}_{m,n}$. Parameter $Z_{n,i}$ denotes the ith frequency domain output data segment at nth output branch. After passing through IFFT and buffered de-interleaving units, final output z_n can be constructed.

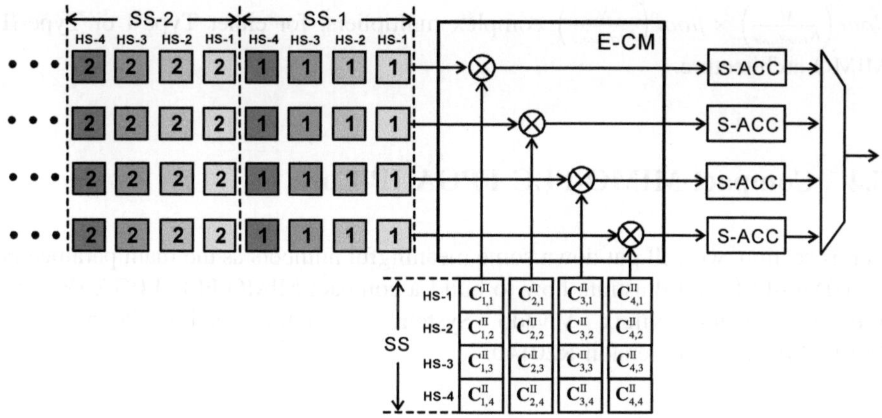

Fig. 5.20 Joint time-space E-CM processing for Type-II MMO

For Type-II MIMO configuration, Fig. 5.20 shows the joint time-space data arrangement using Type-II buffered and de-interleaving procedure.

Similarly, the above processing can also be represented in the matrix format, as Eq. 5.22, where we will do the aggregation along the row. Since the row element represents the data segment, different row element will be corresponding to different data segment at different time slot. Then the aggregation along the row within one SS is actually equivalent to the segment-level accumulator (S-ACC) operation, which summates the HS data segments every SS period.

$$
\begin{aligned}
&\sum^{\text{row}} \begin{bmatrix} X_{1,1} \odot \mathbf{C}^{\text{II}}_{1,1} & X_{2,1} \odot \mathbf{C}^{\text{II}}_{1,2} & X_{3,1} \odot \mathbf{C}^{\text{II}}_{1,3} & X_{4,1} \odot \mathbf{C}^{\text{II}}_{1,4} \\ X_{1,1} \odot \mathbf{C}^{\text{II}}_{1,2} & X_{2,1} \odot \mathbf{C}^{\text{II}}_{2,2} & X_{3,1} \odot \mathbf{C}^{\text{II}}_{3,2} & X_{4,1} \odot \mathbf{C}^{\text{II}}_{4,2} \\ X_{1,1} \odot \mathbf{C}^{\text{II}}_{1,3} & X_{2,1} \odot \mathbf{C}^{\text{II}}_{2,3} & X_{3,1} \odot \mathbf{C}^{\text{II}}_{3,3} & X_{4,1} \odot \mathbf{C}^{\text{II}}_{4,3} \\ X_{1,1} \odot \mathbf{C}^{\text{II}}_{1,4} & X_{2,1} \odot \mathbf{C}^{\text{II}}_{2,4} & X_{3,1} \odot \mathbf{C}^{\text{II}}_{3,4} & X_{4,1} \odot \mathbf{C}^{\text{II}}_{4,4} \end{bmatrix} \\
&= \left[\sum_{n=1}^{4} \left(X_{n,1} \odot \mathbf{C}^{\text{II}}_{1,n} \right) \quad \sum_{n=1}^{4} \left(X_{n,2} \odot \mathbf{C}^{\text{II}}_{2,n} \right) \quad \sum_{n=1}^{4} \left(X_{n,2} \odot \mathbf{C}^{\text{II}}_{1,n} \right) \quad \sum_{n=1}^{4} \left(X_{n,2} \odot \mathbf{C}^{\text{II}}_{1,n} \right) \right]^{T} \\
&= \left[\sum_{n=1}^{4} (Y_{1,n}) \quad \sum_{n=1}^{4} (Y_{2,n}) \quad \sum_{n=1}^{4} (Y_{3,n}) \quad \sum_{n=1}^{4} (Y_{4,n}) \right]^{T} \\
&= [Z_{1,1} \quad Z_{2,1} \quad Z_{3,1} \quad Z_{4,1}]^{T}
\end{aligned} \tag{5.22}
$$

Similarly, after passing through IFFT and buffered de-interleaving units, final output z_n can be constructed.

By doing this effective 2D signal processing, we can actually process multiple E-CM units with fewer physical resources. For example, if $M \times N$ E-CM units are required originally, by the help of 2D data arrangement together with buffered interleaving and de-interleaving techniques, we can implement the E-CM stage (with equivalent multiple E-CM units) of MIMO-FLC algorithm by utilizing only

$floor\left(\frac{M}{N_{Interleave}}\right) \times floor\left(\frac{N}{N_{Interleave}}\right)$ complex multipliers, for either Type-I or Type-II MIMO architecture.

5.4 Compact MIMO-FLC FPGA IP-Core

In this section, we will put down some meaningful numbers as the main parameters of MIMO-FLC and illustrate how to build a compact MIMO-FLC FPGA IP-core. We were using Xilinx Vivado System Generator version 2015.1 and MATLAB/Simulink version R2014b.

5.4.1 Main Parameters and Building Elements

Without loss of generalities, the main common parameters of a typical FPGA-based digital signal processing IP-core include three parts: (1) how fast of the IP-core will be running, (2) how fast of the effective samples will be updated and (3) how accurate of the signal to be processed. More specifically, the above three parameters are corresponding to FPGA processing rate, data sampling rate and input/output signal resolution (i.e., bit-width), respectively. For the MIMO-FLC IP-core, we will need some other parameters, which are given in Table 5.5.

Table 5.5 Exampled values settings for the main parameters of MIMO-FLC IP-core

	Value	Note
CLK_{DATA}	30.72 MHz	Data sampling rate
CLK_{FPGA}	368.64 MHz	FPGA processing rate
N_{Coeff}	160	Number of convolution coefficients
N_{FFT}	512	Size of FFT/IFFT
M	8	Number of inputs
N	8	Number of outputs
B_{width}	16	Input and output signal bit-width
$N_{Interleave}$	8	Calculated according to Eq. 5.20 $= floor\left(\frac{512-160}{512} \cdot \frac{368.64}{30.72}\right) = 8$

Table 5.6 Key elements used for building MIMO-FLC IP-core

Type	Element name	Description
Memory	Single port and dual port RAM FIFO	On-chip block RAM and distributed RAM
DSP unit	DSP48 Marco 3.0	Hardware multiplier
	FFT 9.0	Fast Fourier transform IP

Besides registers and basic elements, like delay, inverter, logical relationship, counter, we need some other elements for building a MIMO-FLC IP-core, those elements are shown in Table 5.6,

5.4.2 Data Path and Coefficient Path

As introduced early in this chapter, FLC processing contains two processing paths, namely, data path and coefficient path. Data path of MIMO-FLC IP-core will be real-time processing because of its nature to handle the continuous incoming data streams, while coefficient path can be pseudo real-time or non real-time. Pseudo real-time scenario is corresponding to fast adaptive coefficients path, in other words, the convolution coefficients will be updated periodically in a short time frame. Non real-time scenario refers to slow adaptive coefficient path, in which the convolution coefficients could be updated in an on-demand way or even in a static manner, like pre-defined as default state.

For the sake of simplicity, here in this section, we will focus on the on-demand coefficient updating scenario, and we assume the time domain convolution coefficients have been pre-converted to the frequency domain and ready to use. Actually in this scenario, i.e., non real-time coefficients updating situation, assigning hardware resources, like hardware FFT core to the coefficient path would be quite a waste, especially in the resource constrained situation. Luckily, the emerging system-on-chip (SoC) combined FPGA chip, like Xilinx Zynq device, would be a very good solution for better trading off the software and hardware resources. Let's still use the on-demand coefficients updating scenario as an example, software-based FFT can be used in this situation, and after fulfilling the purpose of FFT, the corresponding computational resource (CPU time) can be released for other applications. This software-defined approach is quite suitable for the modern complex digital systems, especially the software-defined wireless systems.

5.4.3 System Top Module

As discussed in Sect. 5.3, two MIMO-FLC processing approaches are corresponding to two different MIMO types, i.e., Type-I and Type-II. Similarly, we will have two MIMO-FLC IP-core architectures with slightly difference. Figures 5.21 and 5.22 illustrate the top modules of Type-I MIMO-FLC base core and Type-II MIMO-FLC base core, respectively. Both of them contain six processing stages. Two out of those six stages, namely input buffer stage and output buffer stage are exactly the same, while the other four stages of two base cores are slightly different comparing to their counterparts. This is because the architectural differences between Type-I MIMO-FLC and Type-II MIMO-FLC regarding the arrangement of coefficients. And we'd like to get rid of the redundant processing as much as

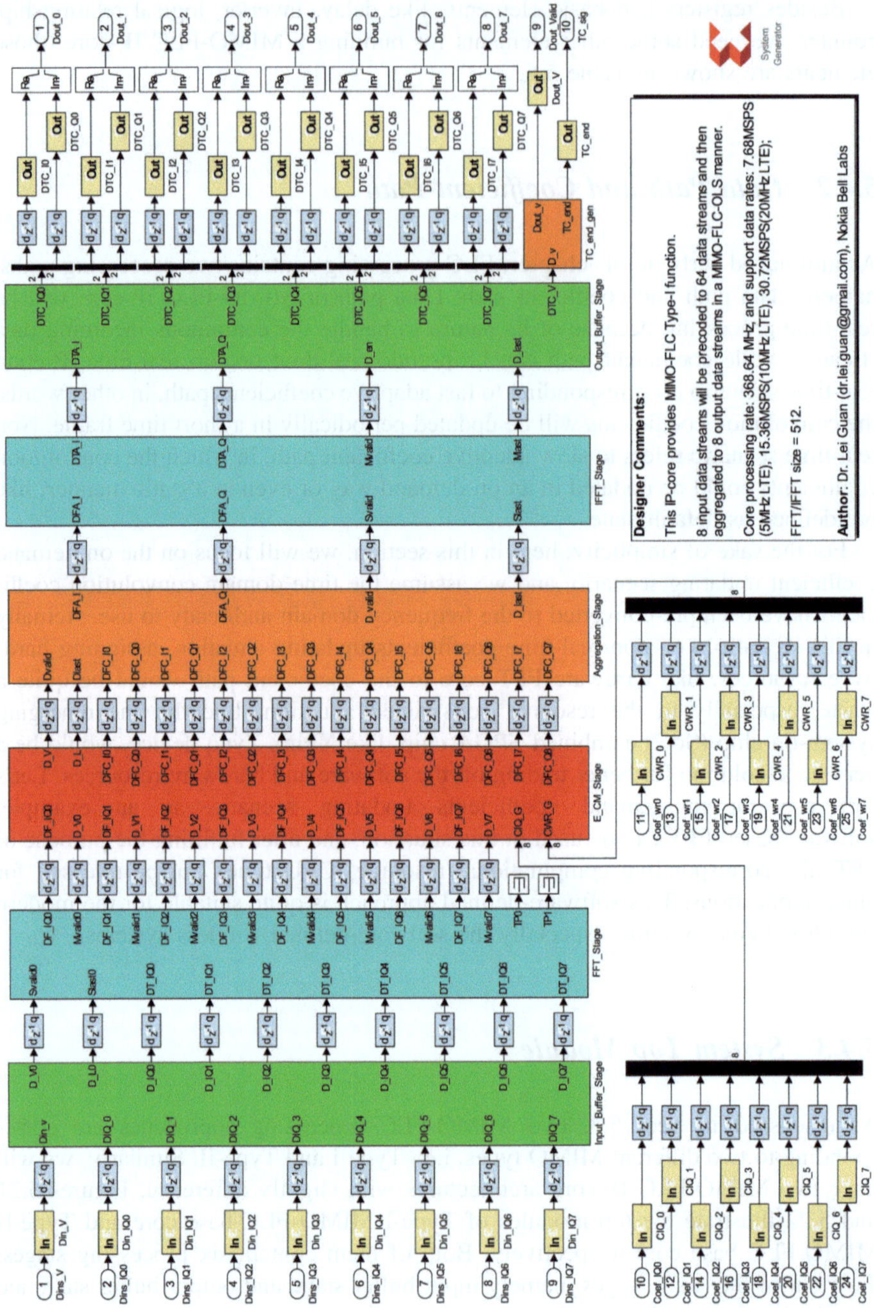

Fig. 5.21 Top module of MIMO-FLC IP-core (Type-I MIMO-FLC)

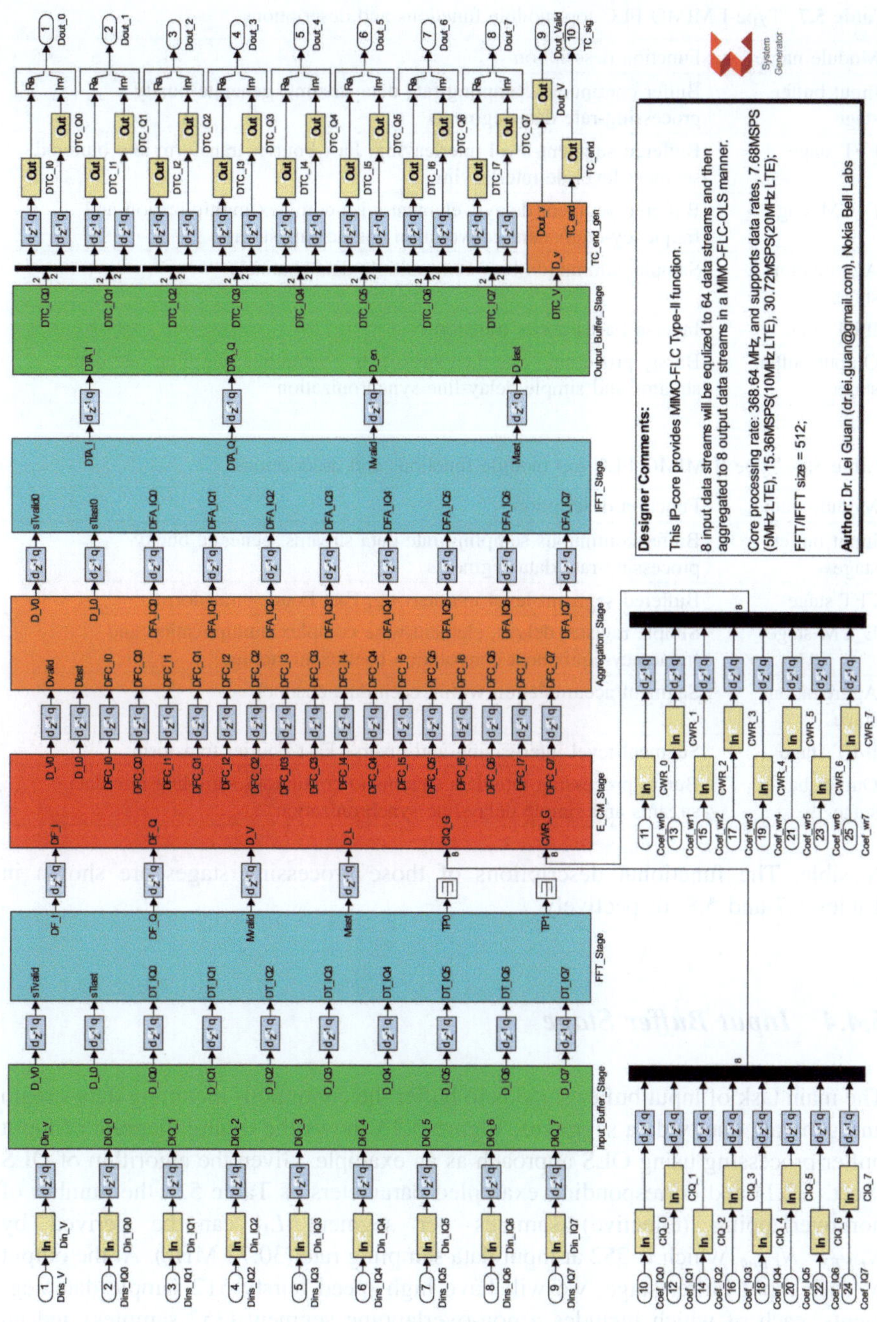

Fig. 5.22 Top module of MIMO-FLC IP-core (Type-II MIMO-FLC)

Table 5.7 Type-I MIMO-FLC top module functions and descriptions

Module name	Function description
Input buffer stage	Buffer continuous sampling-rate data streams, generate bursty processing-rate data segments
FFT stage	Buffered segment-level interleaving, Fast Fourier transform and buffered segment-level de-interleaving
E_CM stage	Buffered segment delays, element-wise complex multiplication and frequency-equivalent convolution coefficients storage
Aggregation stage	Sample summation between multiple data-channels
IFFT stage	Inverse Fast Fourier transform
Output buffer stage	Bursty processing-rate data streams to continuous sampling-rate data streams and simple delay-line synchronization

Table 5.8 Type-II MIMO-FLC top module functions and descriptions

Module name	Function description
Input buffer stage	Buffer continuous sampling-rate data streams, generate bursty processing-rate data segments
FFT stage	Buffered segment-level interleaving, Fast Fourier transform
E_CM stage	Simple register delays, element-wise complex multiplication and frequency-equivalent convolution coefficient storage
Aggregation stage	Segment accumulation within each data-channel
IFFT stage	Segment-level interleaving and inverse Fast Fourier transform
Output buffer stage	Bursty processing-rate data streams to continuous sampling-rate data streams and simple delay-line synchronization

possible. The functional descriptions of those processing stages are shown in Tables 5.7 and 5.8, respectively.

5.4.4 *Input Buffer Stage*

The main task of input buffer stage is to buffer the continuous incoming data stream and generate bursty data segments. Figure 5.23 shows the timing diagram of input buffer processing using OLS approach as an example. Given the algorithm of OLS (Sect. 5.2.1) and corresponding exampled parameters as Table 5.5, the number of non-overlapping (effective) samples per segment L_D can be derived by $N_{FFT} - N_{Coeff}$, which is 352 at input data sampling rate (30.72 MHz). At the output of each input buffer stage, we will have high-speed bursty 512-sample data segments, each of which includes a non-overlapping segment (352 samples) and an overlapping segment (160 samples). Within the period of one data segment at data sampling rate, i.e., 352 × 1/30.72 ms, 4224 (352 × 368.64/30.72) samples at processing rate can be processed. More specifically, because of 4224 = (512 + 16) × 8,

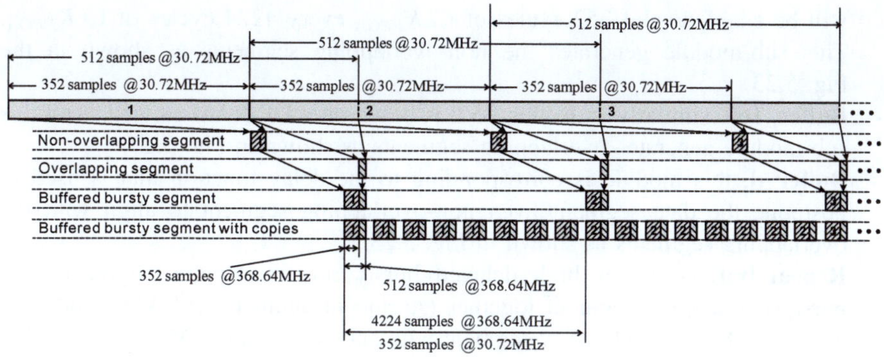

Fig. 5.23 Timing diagram of input buffer processing for one branch

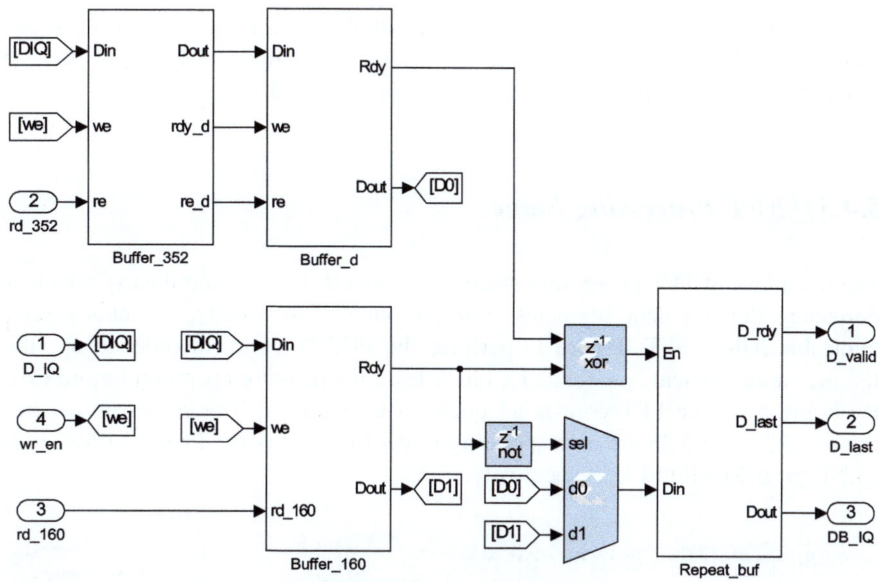

Fig. 5.24 Sub-modules of one input buffer at input buffer stage

within the period of one data segment at sampling rate), 8 high-speed bursty 512-sample data segments can be fitted with some guard period (16 samples at processing rate) between two adjacent high-speed data segments.

The input buffer stage of the MIMO-FLC base core contains 8 parallel input buffer units, each of which is illustrated in Fig. 5.24.

Each one of input buffer units contains the following sub-modules:

– **Buffer_352**, is based on a DP-RAM. The continuous input data are written to one port of the DP-RAM every 12 cycles of CLK_{FPGA} continuously (equivalent to every cycle of CLK_{DATA}). While the other port of the DP-RAM, i.e., read port will be enabled/valid 352 cycles of CLK_{FPGA} every 4224 cycles of CLK_{FPGA}. This sub-module generates the non-overlapping segments as shown in the Fig. 5.23.

- **Buffer_160**, similarly as Buffer_352, is based on a DP-RAM as well. And this sub-module generates overlapping segments as shown in Fig. 5.23.
- **Buffer_d**, this module is corresponding to the delay function that is used to postpone the data segment from non-overlapping segment to align with the overlapping segments as shown in Fig. 5.23.
- **Repeat_buf**, combines both delayed bursty non-overlapping segments and bursty overlapping segment together (storing in another DP-RAM), and then generates the buffered bursty segment with another 7 copies. More specifically, updating the contents of DP-RAM (by writing to one port) every 4224 cycles, and pushing the contents out of DP-RAM every 512 + 16 = 528 cycles of CLK_{FPGA}.

The outputs of input buffer stage are 8 parallel complex data branches and two control signals which are indicating the data valid period (512-sample per 528 samples period) and the last sample of each valid period.

5.4.5 FFT Processing Stage

The main task of FFT processing stage is to convert time domain data segments to frequency domain data segments. For the sake of simplicity, we choose customizable Xilinx FFT IP-core to perform the FFT function. In order to use this IP-core more efficient, we created a block level interleaving operation before FFT. By doing this, one FFT core is adequate to take care of 8 branches processing. Figures 5.25 and 5.26 are corresponding to the FFT stages for Type-I MIMO-FLC and Type-II MIMO-FLC, respectively.

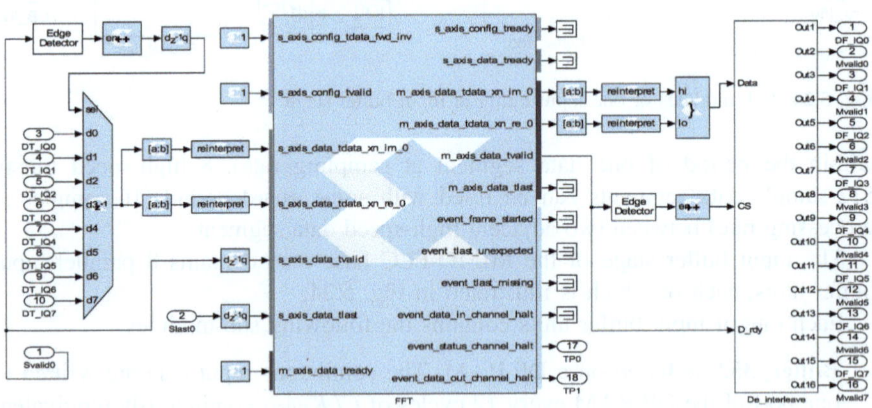

Fig. 5.25 FFT stage of Type-I MIMO-FLC

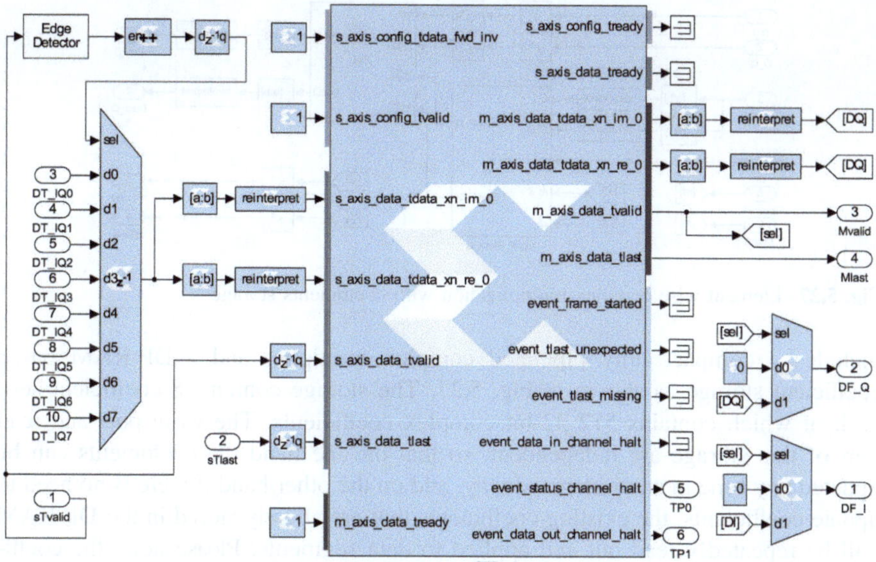

Fig. 5.26 FFT stage of Type-II MIMO-FLC

At the output of Type-I MIMO-FLC FFT stage, there is block level de-interleaving module, which de-interleaves the frequency domain data segments into 8 different parallel data paths. While at the Type-II MIMO-FLC FFT stage, the output is interleaved data segments which will splits into 8 same parallel data paths. Those arrangements will generate the corresponding 2D data array as illustrated in Figs. 5.19 and 5.20.

To maximize the throughput of FFT processing, the core is customized as pipelined streaming I/O as the module interfaces, which allows continuous data processing. In our example, the FFT core will process the segment-level interleaved input data and generate corresponding 512-sample segment with 16-sample guard gap between the adjacent two segments. Moreover the Xilinx FFT IP-core can be actually customized for different design targets. Specifically, the butterfly processing and internal storages can be configured using delicate resources like DSP48s and BRAMs or using slices and distributed memories (e.g., LUTs and FFs). Again it is a trade-off game between different requirements given physical constrains. More information about this IP-core is available from Xilinx Website.

5.4.6 E-CM Processing Stage

The main task of E-CM processing stage is to perform element-wise complex multiplication between data streams and corresponding frequency coefficients. In our example, E-CM stage contains 8 parallel E-CM sub-modules, each of which

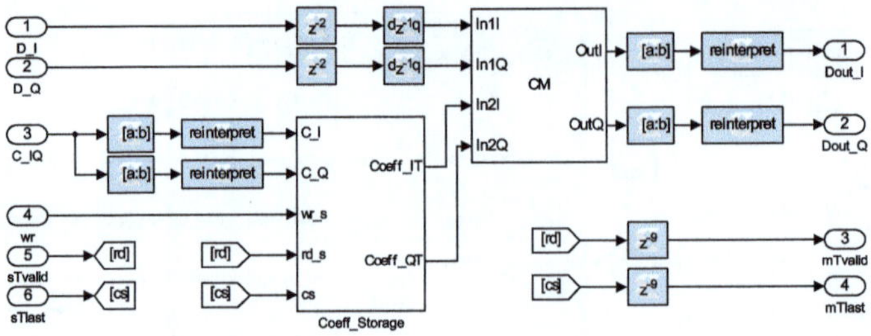

Fig. 5.27 Element-wise complex multiplication with coefficients storage

includes a compact fully pipelined complex multiplier and a DP-RAM based coefficient storage as shown in Fig. 5.27. The storage contains 8 coefficient sets, each of which contains 512 32-bit complex coefficients. The write port and read port of this storage are independent, so that on one hand the coefficients can be updated any time in an on-demand way, and on the other hand if there is no need to update coefficients, the existing coefficients that are already stored in the DP-RAM will be repeatedly read out and applied to data segments. Please note, the coefficients should be pushed out in the same timing relationship as bursty data path.

Figure 5.28 illustrates the compact complex multiplier architecture using 3 DSP48 macros with 6 clock cycles latency. A direct architecture of complex multiplier can be used as well with 4 DSP48 macros.

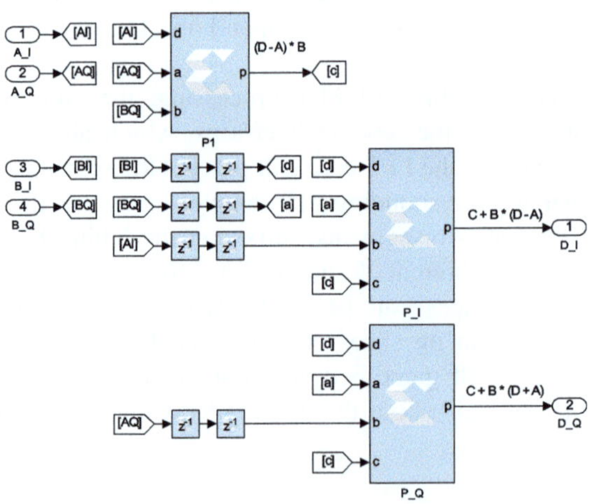

Fig. 5.28 3-DSP48 fully pipelined complex multiplier implementation architecture

5.4.7 Aggregation Stage

Depending on the MIMO type, two different aggregation stages are required. For Type-I MIMO-FLC, according to the processing flow shown in Fig. 5.19, a simple tree-structure addition based aggregation can be used to aggregate the data segments between multiple branches. A straightforward implementation of this aggregation stage is illustrated in Fig. 5.29.

While for Type-II MIMO-FLC, according to the processing flow shown in Fig. 5.20, we need segment accumulation based aggregation to aggregate the 8 relevant data segments among multiple segments within one branch. And since there are 8 parallel E-CM outputs, we require 8 segment accumulation units. A simple way to achieve segment accumulation is designed as shown in Fig. 5.30. Extended from simple accumulator, the most effective architecture of segment accumulation is to use a DP-RAM after addition operation. The DP-RAM temporally stores the addition results (write operation) and pushes out (read operation) the results to participate the next segment addition until reaching to the end of 8 segments. And the end of 8 segments cycle, i.e., 4224 samples period, the accumulated results will be latched and pushed out of the E-CM stage.

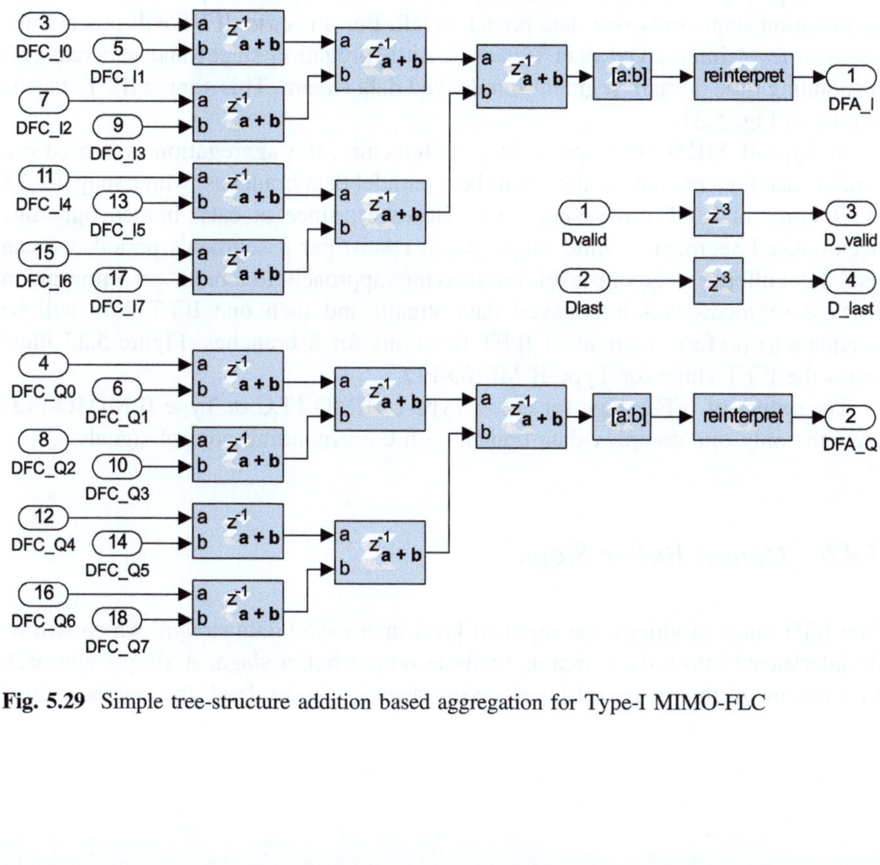

Fig. 5.29 Simple tree-structure addition based aggregation for Type-I MIMO-FLC

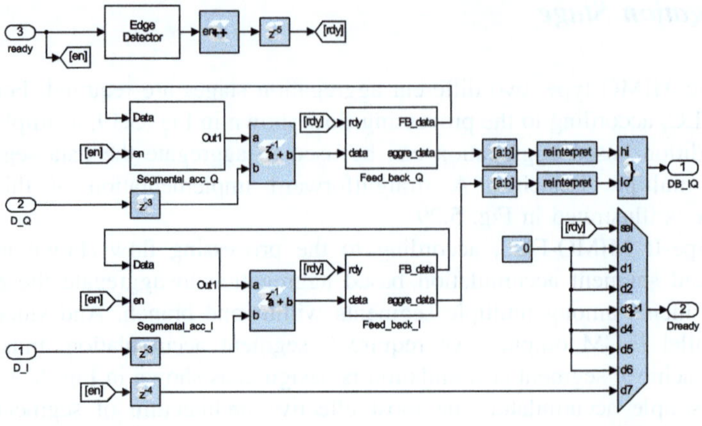

Fig. 5.30 Segment accumulation based aggregation for Type-II MIMO-FLC

5.4.8 IFFT Processing Stage

The main task of IFFT processing stage is to convert the frequency domain processed data back to the time domain. IFFT function can be easily achieved by Xilinx FFT IP-core, i.e., only one control bit needs to be configured.

In Type-I MIMO-FLC processing architecture, after simple addition based aggregation stage, only one data branch is left. For this case, IFFT will process this segment-level interleaved data stream (from aggregation stage) and generate corresponding time domain segment-interleaved data stream. This simple IFFT stage is shown in Fig. 5.31.

In Type-II MIMO-FLC processing architecture, the aggregation is carried out within each data branch; so there will be 8 parallel data branches, which may need 8 IFFT cores at IFFT processing stage. However, since at each branch only one accumulated segment is valid (aggregation result) per 8 segments period, we can use the buffered segment-level interleaving approach to create an information lossless segment-level interleaved data stream and then one IFFT core will be adequate to perform equivalent IFFT functions for 8 branches. Figure 5.32 illustrates the IFFT stage for Type-II MIMO-FLC.

The output of IFFT stage for either Type-I MIMO-FLC or Type-II MIMO-FLC contains only one complex data branch with corresponding control signals.

5.4.9 Output Buffer Stage

The IFFT stage produces one segment-level interleaved data stream, which will be de-interleaved into 8 data streams firstly at output buffer stage. A simple approach to implement this segment-level de-interleaving is to feed the segment-level

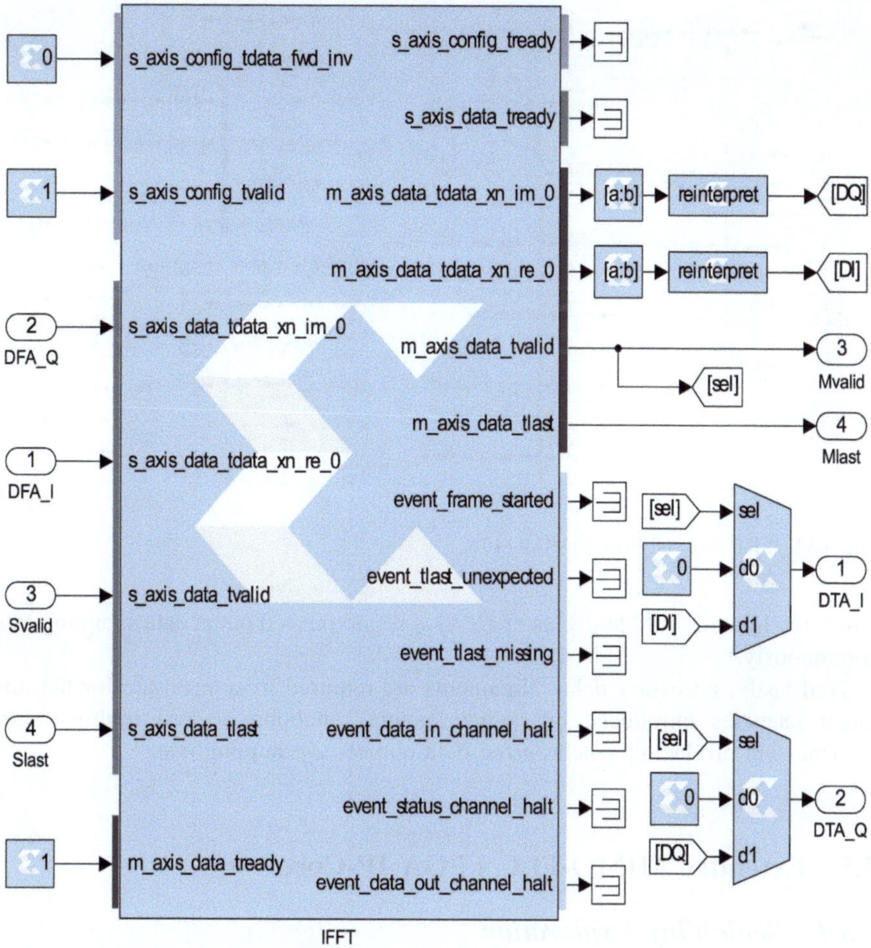

Fig. 5.31 IFFT stage of Type-I MIMO-FLC

interleaved data stream to 8 channel selection module (CH_x module in the Fig. 5.33, x = 1, 2, …, 8), each of which will use a segment-enable signal to pick up the particular segment every 8 segments cycle.

Secondly, at each data channel, a DP-RAM module is used to buffer the bursty processing rate data segment to continuous sampling rate data stream. For the OLS based FLC processing, the first N_{Coeff} samples of each segment will be discarded, while for the OLA based FLC processing, the last N_{Coeff} samples of each segment will be added with the first N_{Coeff} samples of the next segment. In our example OLS approach, the first 160 samples of each data segment will be discarded straightforwardly, and the rest 352 samples of each data segment will be written to the

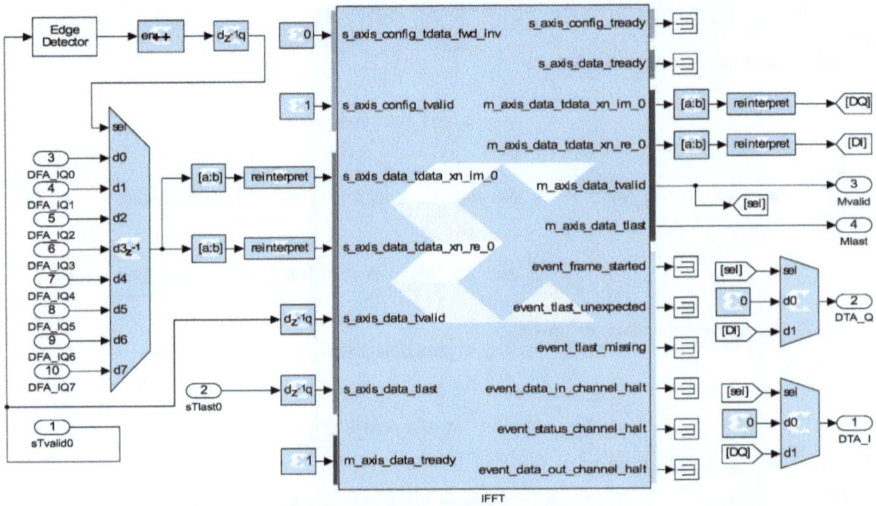

Fig. 5.32 IFFT stage of Type-II MIMO-FLC

buffer module every 4224 cycles of CLK_{FPGA}, and pushed out at data sampling rate continuously.

And lastly, necessary delay alignments are required to compensate for the different latencies introduced by de-interleaving functions, so that multiple data branches are producing synchronized data outputs at sampling rate.

5.5 Extended MIMO-FLC FPGA IP-Core

5.5.1 Scalability Exploration

In order to build a scaled MIMO application, we can use two approaches. The first one is to directly stack multiple standard MIMO-FLC base-cores in parallel. The second approach is to create an extended IP-core by combining multiple standard MIMO-FLC base-cores, which shares partial functional units. For example, Figs. 5.34 and 5.35 illustrate the two approaches to build 8-to-16 Type-I MIMO and 16-to-8 Type-II MIMO, respectively.

By using extended architecture, we can efficiently create scaled MIMO-FLC application, and the larger size of the application, the average resource utilization will be. However, MIMO-FLC IP-core is a relatively large DSP system design on FPGA, building a single larger scaled core may not be the optimized solution regarding the real FPGA timing performance. Again, it is a trade-off game, and reasonably integrating more units in one logic core would be the “best” solution considering both FPGA resources and actual timing performance.

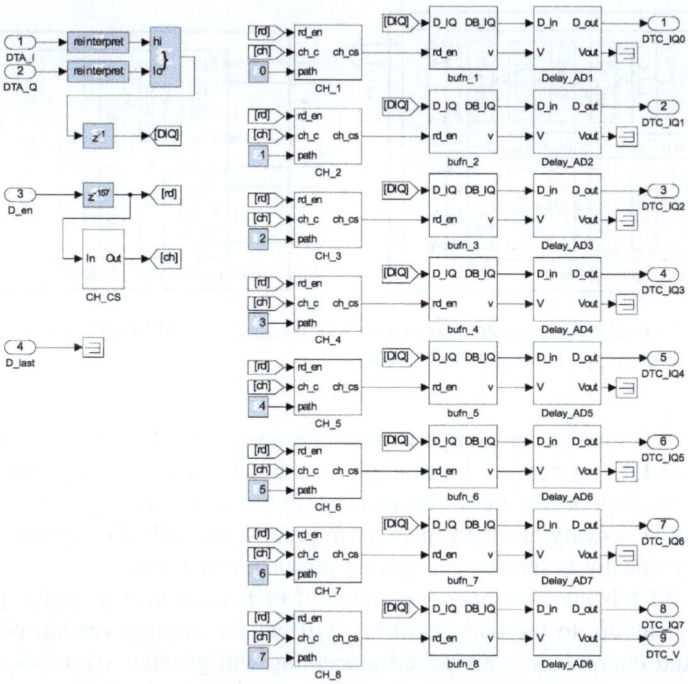

Fig. 5.33 Output buffer stage of MIMO-FLC base core

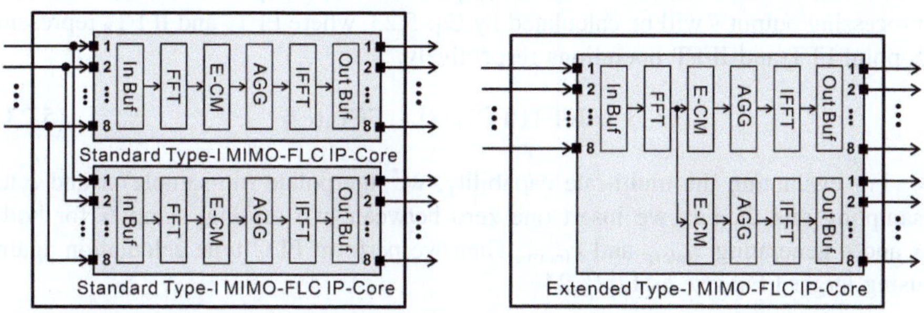

Fig. 5.34 8-to-16 Type-I MIMO using two 8-to-8 standard Type-I MIMO-FLC IP-cores or one 8-to-16 extended Type-I MIMO-FLC IP-core

5.5.2 Flexibility Exploration

Multi-rate single core operation is always a beneficial approach regarding system flexibility. Specifically, for maximizing the system flexibility or reserving the processing headroom for future system upgrading, wireless engineers are always

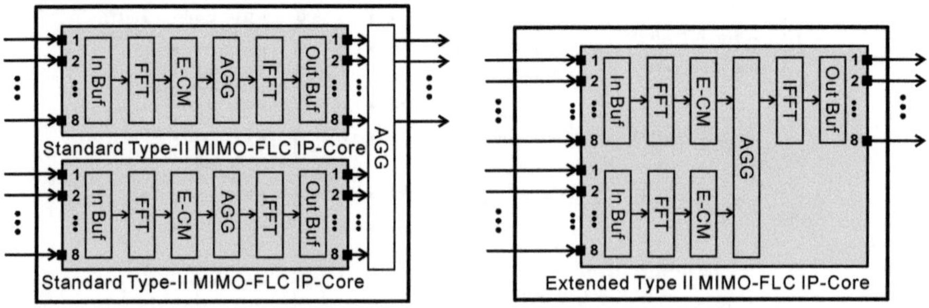

Fig. 5.35 16-to-8 Type-II MIMO using two 8-to-8 standard Type-II MIMO-FLC IP-cores or one 16-to-8 extended Type-II MIMO-FLC IP-core

looking for a solution can support multi-rate operations. Using 4G wireless communication LTE system as an example, the baseband processing unit is usually designed for supporting multi-rate scenarios, i.e., the processing unit designed for 20 MHz LTE usually has the capability to process 10 MH LTE signals, unless the unit is specifically designed with certain resource constraints.

Since IFFT is inversed linear operation of FFT, theoretically applying FFT and IFFT sequentially to the data segments will get the original version of data segments. And interpolating this paired processing will generate corresponding interpolated outputs. Now, we will use a quite simple and straightforward approach to evaluate the multi-rate processing capability of FLC processing.

Assume we have a data segment x with 8 complex number samples at sampling rate f and a coefficient set c with 8 complex number samples, then the FLC-type processing output y will be calculated by Eq. 5.23, where FFT_8 and IFFT_8 represent 8-point FFT and IFFT operations respectively.

$$y = \mathrm{IFFT}_8[\mathrm{FFT}_8(x) \odot \mathrm{FFT}_8(c)] \tag{5.23}$$

For evaluating the multi-rate capability, we interpolate (up-sample) x and c to sampling rate $2f$, i.e., we insert one zero between two original samples for both x and c generating x_{Interp} and c_{Inerp}. Then we perform FLC type calculation again using larger FFT size as Eq. 5.24.

$$y_{\mathrm{Interp}} = \mathrm{IFFT}_{16}\left[\mathrm{FFT}_{16}(x_{\mathrm{Interp}}) \odot \mathrm{FFT}_{16}(c_{\mathrm{Interp}})\right] \tag{5.24}$$

Figure 5.36 illustrates the above processing with random number data and coefficient samples. Comparing y with y_{Interp}, it is not difficult to realize that they contain exactly the same meaningful information. This verification gives a sense that, if we design a FPGA FLC core for data sampling rate CLK_{DATA}, and we can directly use this core to perform FLC processing for CLK_{DATA}/N sampling rate application (N is an integer) by simply interpolating the input data and coefficients, then the FPGA FLC core will be "fooled" by this simple interpolation operation and carry out the digital convolution between the interpolated data and interpolated

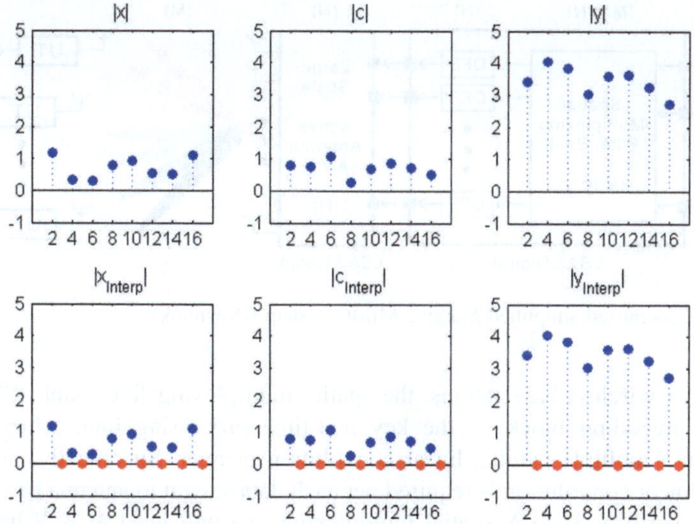

Fig. 5.36 Multi-rate scenario illustration with interpolation

coefficients in a desired way. Let's think of it in another way, FLC is functionally equivalent to linear convolution, which is multi-rate processing approach. In other words, if we treat the linear convolution and FLC as black boxes, because of their equivalent functionality, FLC will have the same feature as linear convolution. In other words, we can say, if we design a MIMO-FLC for 20 MHz LTE application, it will be valid for 10 MHz and 5 MHz LTE application as well.

5.6 Applications in Wireless Communications

Large scaled antenna system (LSAS) or massive MIMO system has been actively indentified as one of the promising candidates for the forthcoming fifth generation (5G) wireless communications [7]. LSAS will utilize software-defined open-loop steerable beamforming and closed-loop MIMO techniques to spatial multiplex data streams for multiple users [1]. For illustration, a simplified diagram of LSAS based future wireless base-station is shown in Fig. 5.37 including two main parts: (1) LSAS digital part for dynamic data streams spatial multiplexing (from M input data streams to N output data streams) and standard N-branch digital front end (DFE) processing, and (2) LSAS radio part for converting the N-branch digital baseband signals to full-size N-branch radio frequency (RF) signals and radiating the signals to the free space by N antenna elements. UT represents universal terminal, which could be small cell, super house user terminal or handset-like user equipment.

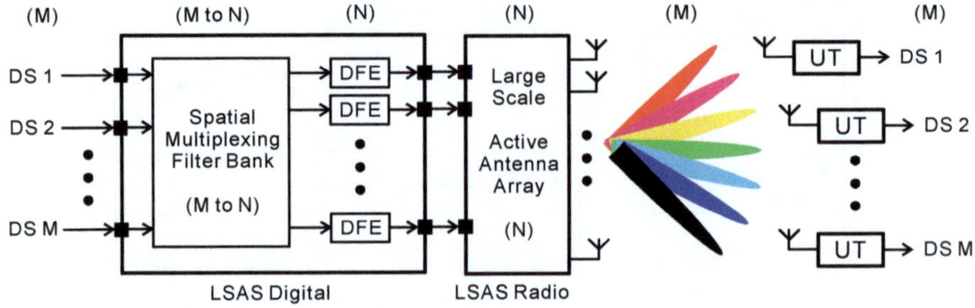

Fig. 5.37 Generalized simplified Massive MIMO system (Downlink)

In LSAS wireless basestations, the spatial multiplexing filter bank (SMFB) or so-called precoding block, is the key real-time processing unit, which can be realized in a multiple parallel linear convolution manner. Specifically, one digital complex linear convolution is required per each data stream to antenna pair, in other words, to perform M to N spatial multiplexing, we will need $M \times N$ linear convolutions and N addition operations. Architecturally, linear convolution shares the same structure as finite impulse response (FIR) filter, which are effectively coefficients weighted multiplication and addition operations. However, though it is quite simple, FIR filter based architecture is not area-efficient approach to construct compact SMFB on the FPGA platform, especially when long-tap asymmetrical wireless-channel related coefficients are required, because of its requirement for a large amount of hardware multiplier resources. An alternative solution is the MIMO-FLC approach, which will be discussed in Sect. 5.6.1.

Another promising application of MIMO-FLC IP-core is for building a baseband equivalent MIMO channel emulator. Wireless channel emulators are a type of RF devices to mimic different transmission condition for testing air interface. Baseband equivalent channel emulator uses mathematical models together with its digital implementation to perform the equivalent analog channel effects in the real medium. Fading (both time domain and frequency domain) effects are the main characteristic of a typical channel emulator. The digital convolution is one of the most important approaches to build a baseband equivalent channel emulator, and thus the MIMO-FLC IP-core plays a very important role in this application.

5.6.1 *Spatial Multiplexing Filter Bank for Massive MIMO System*

In this section, in order to proof the concept for real wireless applications, we created several MIMO-FLC on Xilinx FPGAs with different configurations for comparison as shown in Table 5.9. You will have a sense how much key DSP48 resources can be saved by using MIMO-FLC base cores comparing to the compact semi-parallel FIR implementation architecture.

Table 5.10 shows the FPGA resource utilization for different MIMO-FLC base cores, comparing to the FIR base architecture, clearly, the MIMO-FLC can significantly reduce the number of the hardware multipliers (DSP48s) for the same digital convolution functions.

The number of required DSP48s for compact semi-parallel FIR-base linear convolution architecture [8] can be calculated as Eq. 5.25, where we assume the compact 3-DSP48 architecture is used for building a complex multiplier. Function $ceil(\cdot)$ derives the next larger integer of the value.

$$N_{\text{SP-LC}}^{\text{HM}} = ceil\left(3 \cdot N_{Coeff} \cdot M \cdot N \cdot \frac{CLK_{FPGA}}{CLK_{DATA}}\right) \quad (5.25)$$

Scaled SMFB can be straightforwardly constructed by stacking multiple MIMO-FLC base cores together with some simple aggregation units at the output stages. However, as we discussed before, in order to better use the FPGA resources

Table 5.9 The parameters of MIMO-FLC base cores

	Base core 1	Base core 2	Base core 3	Note
CLK_{DATA}	30.72 MHz	30.72 MHz	30.72 MHz	Data sampling rate
CLK_{FPGA}	307.2 MHz	491.52 MHz	368.64 MHz	FPGA processing rate
N_{Coeff}	128	116	160	Number of convolution coefficients
N_{FFT}	256	512	512	Size of FFT and IFFT
M	1	1	8	Number of inputs
N	4	12	8	Number of outputs
B_{width}	16	16	16	bit-width of input and output signals
$N_{Interleave}$	4	12	8	Calculated according to Eq. 5.20

Table 5.10 FPGA resource utilization of different MIMO-FLC base cores

	Base core 1	Base core 2	Base core 3	Total
DSP48s[1] (DPS48s)[2]	37 (154)	29 (261)	129 (2560)	3600
BRAMs	19	54	142	2940
LUTs	6819	9639	13,347	433,200
FFs	13,402	12,827	25,616	866,400

[1]Depending on the Xilinx FFT IP-core setting, the actual utilization of DSP48s for the same function would be a little bit different. Please refer to the Xilinx FFT IP-core data sheet

[2]Reference DSP48s utilization if using compact semi-parallel FIR-based linear convolution implementation architecture

when the number of antennas are scaling up, we can extend the MIMO-FLC standard base core.

Then we implemented several SMFBs using extended MIMO-FLC base core 3 for both transmitter (Type-I MIMO-FLC) and receiver (Type-II MIMO-FLC) for different scaled number of antennas (N_{ANT} = 8, 16, 32, 64) on Xilinx Ultrascale KU115 device. The utilization percentages of the key resources are shown in Fig. 5.38. Particularly, in the 64 antenna scenario, the utilization of BRAMs and DSP48s are less than 50% and 30%, respectively, leaving quite a large percentage of resources for other necessary functions in the LSAS base-station, such as digital front end and RF signal conditioning functions, like digital predistortion [9].

Furthermore, as shown in Fig. 5.39, the resource utilization of 8-antenna scenario is normalized to one, and then we calculate the normalized resource utilization per 8-antenna for different scaled scenarios. The average resource utilization is getting lower when the number of antennas is scaled up. In other words, under certain physical constrains, the higher the number of antennas is in the massive MIMO system, the lower the average digital resource utilization for SMFB.

As validated in last section, MIMO-FLC architecturally supports multi-rate application. For example, in the case study, the MIMO-FLC core is customized for 20 MHz LTE (30.72 Msps) application, without changing anything, it can be directly used for 10 MHz LTE (15.36 Msps) application by simply interpolating the signal by one zero every single sample. While the frequency dependant convolution coefficients for 10 MHz LTE will be proportionally reduced half of the length as that for 20 MHz application, so similar simple interpolation needs to be in place for generating FLC coefficients as well.

Cost-effective large scale antenna system will be in high demand for the next generation wireless network. Compact MIMO-FLC can significantly reduce the number of hardware multipliers for constructing SMFB, thus reducing the burden

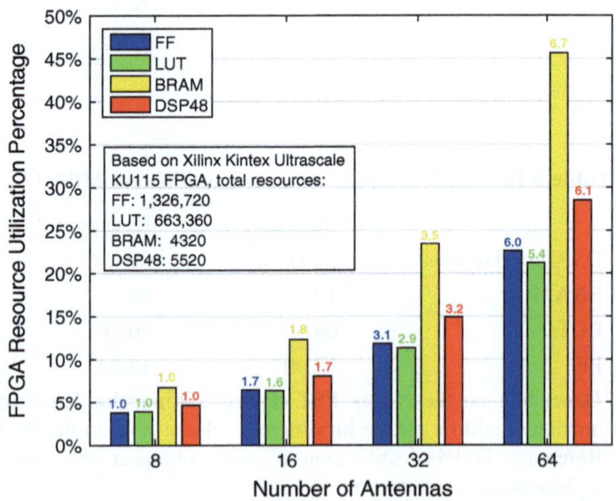

Fig. 5.38 MIMO-FLC complexity performance in scaled SMFB application

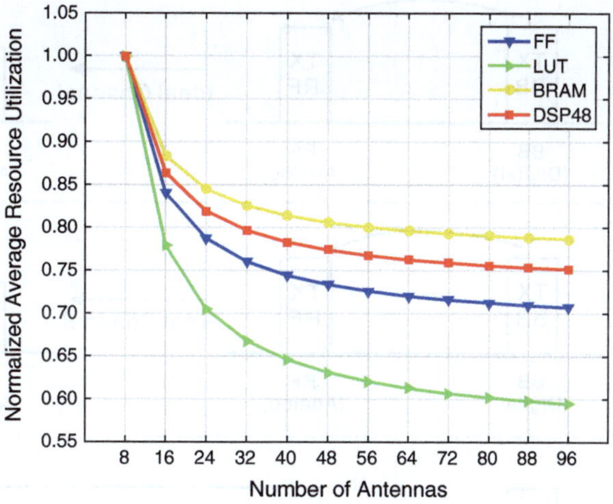

Fig. 5.39 Normalized average resource utilization for scaled up antennas

of digital processing for Massive MIMO system. MIMO-FLC enables an appealing single-chip LSAS digital solution, which will open a wider door to the real cost-effective software defined 5G infrastructure deployment.

5.6.2 Simplified Baseband Equivalent MIMO Channel Emulator

Channel emulators are normalized used for facilitating the air interface testing in wireless communication. And it can be very useful to validate the baseband processing as well, to see whether the particular processing, such as equalization, can compensate for the channel fading (distortions). The real wireless channel is the air, or the free space. In order to simply the tests with well controlled pre-known channel situation, channel emulator mimics the real channel with some controllable physical medium. And baseband equivalent (BBE) channel emulator use baseband equivalent channel model to mimic the affects happened to the in-band transmitted signal.

For example, Fig. 5.40 illustrates the basic idea of BBE channel emulator in the frequency domain. Let's assume that the RF devices, such as up-convertor, frequency mixer, bandpass RF filter are flat within the interested transmission frequency band, however in the real world, they are not strictly flat in the frequency domain, and some compensation schemes have to be used to flatten those RF imperfections. As shown in Fig. 5.40, baseband TX signal (TX BB) will be upconverted to RF at the transmitter side before radiating to the air, and TX RF signal passing via idea channel (without any fading), will only lose the power (the magnitude). However, if TX RF signal passes via real channel (with frequency dependent fading), the received signal at the receiver side will not only lose the

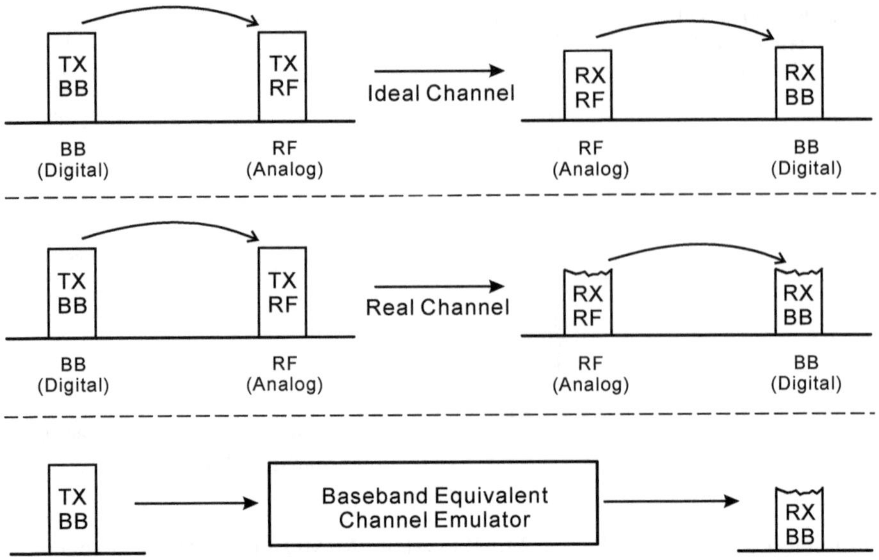

Fig. 5.40 Baseband equivalent channel emulator illustration

power, but also introduce in-band ripple (frequency dependent distortion). And after down conversion, the RX BB signal will have the same frequency dependent distortion due to the channel fading. The baseband equivalent channel emulator actually will significantly simply the channel effect simulation, i.e., putting the RF devices frequency responses and channel fading into consideration together, the input of the BBE channel emulator is the TX BB and the output of BBE channel emulator is the distorted RX BB signal. By doing this, one can easily verify whether the baseband processing algorithms are capability to handle the channel related distortions.

To build a compact BBE channel emulator is quite simple if using MIMO-FLC approaches, for example, as illustrated in Fig. 5.41, two FPGA evaluation boards with an extra PC will be enough to start a MIMO baseband emulation platform. One FPGA is used for doing LSAS digital work using MIMO-FLC cores to construct a scaled SMFB function, and another FPGA board is used for channel emulation, which can also use MIMO-FLC cores with automatic gain control (AGC) function to mimic channel fading. We can create many arbitrary channel situations to thoroughly evaluate the essential Massive MIMO processing, and channel situations (channel fading coefficients) can be derived by real channel measurements as well. This flexible low-cost Massive MIMO baseband verification solution would save quite a bit of R&D time at the early proof of concept stage, and it is also very useful when doing second-round system upgrading with new improved baseband algorithms.

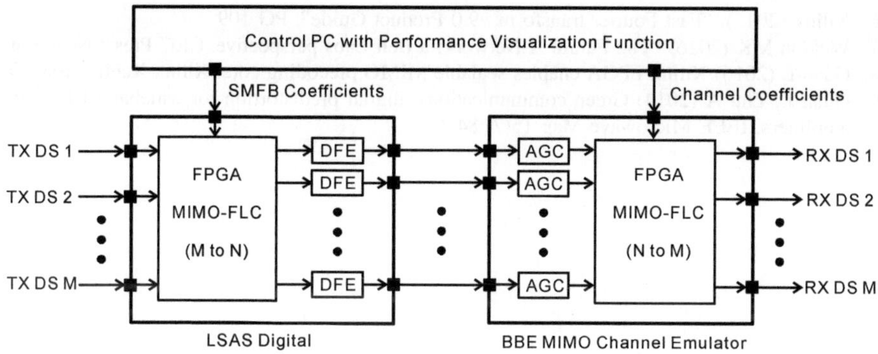

Fig. 5.41 MIMO-FLC IP-core based Massive MIMO emulator including LSAS digital processing unit and BBE MIMO channel emulator

5.7 Conclusions

This chapter introduces compact MIMO-FLC approaches with inside-out necessary information, like FFT basic, FLC basic, overlapping save (OLS) approach, overlapping add (OLA) approach, detailed FPGA MIMO-FLC core implementation architecture, and extended MIMO-FLC IP-core and its appealing applications for the future wireless system, like Massive MIMO system. In general, complexity, scalability, flexibility and expendability are the four main concerns for designing modern wireless system. The ideas and design approaches developed in this chapter are mainly following the progress of a tense R&D project, which was targeting a cost-effective scalable Massive MIMO testing platform for the 5G wireless communications. However, the practices, system design strategies and detailed design tricks are not limited to the MIMO-FLC application, and they can be directly utilized for building other similar type dense FPGA-based digital signal processing modules.

References

1. Marzetta TL (2015) Massive MIMO: an introduction. Bell Labs Technical Journal, pp 11–22
2. E-UTRA (2011) physical channels and modulation, release 10, 3GPP Standard, TS 136.211
3. Guan L (2015) Compact scalable frequency dependent precoding and its FPGA implementation for 5G massive MIMO wireless systems. IET Electr Lett 51(23):1937–1939
4. Cooley JW, Tukey JW (1965) An algorithm for the machine calculation of complex Fourier series. Mathem Comput 19:297–301
5. Burrus CS, Parks TW (1985) DFT/FFT and convolution algorithms and implementation. Wiley, NY

6. Xilinx (2015), "Fast Fourier transform v9.0 Product Guide", PG 109
7. Weldon MK (2016) "The Future X Network: a Bell Labs perspective. CRC Press, New York
8. Guan L (2016) Xilinx FPGA enables scalable MIMO precoding core. Xilinx Xcell J, issue 94
9. Guan L, Zhu A (2014) Green communications: digital predistortion for wideband RF power amplifiers. IEEE Microwave Mag 15(7):84–99

Chapter 6
FPGA-based DSP System Verification

Abstract This chapter will highlight the verification approaches for validating the functionality of FPGA-based DSP systems at both of the system level and the chip level using simulation-based verification approach and hardware-based verification approach, respectively. Examples are given for illustrating the flexible SW and HW co-operated test-bench platform, targeting the validation of DPD-based PA linearization application and MIMO-FLC-based massive MIMO precoding application.

6.1 Introduction

Without any surprise, validation of the designed FPGA-based DSP sub-system is critical and essential regarding the delivery of the designed digital systems in a project. The basic concept of the validation is to verify the designed function against the requirement or specification, simply speaking, the purpose is to check whether the designed logic (digital function) can do the job or not, what's the performance of the design, and how to evaluate the corresponding performance of the design. The best practice is to split the big system (e.g., the MIMO-FLC IP-core in Chap. 5) into smaller logical groups (e.g., input buffers, FFT/IFFT) and verify them against the so-called reference models one by one in terms of the functionality and interfaces between each sub-system. As the beginning of this chapter, Fig. 6.1 illustrates a recommended system design and validation flow-chart, and we will follow the flow chart to tell the story.

6.2 Verification Platforms

For small logic design and verification, FPGA vendor tools are very helpful, however if we need to design and validate a relative complex DSP system, e.g., MB-FLC, Matlab/Simulink based approaches would be a better choice from

L. Guan, *FPGA-based Digital Convolution for Wireless Applications*,
Springer Series in Wireless Technology, DOI 10.1007/978-3-319-52000-1_6

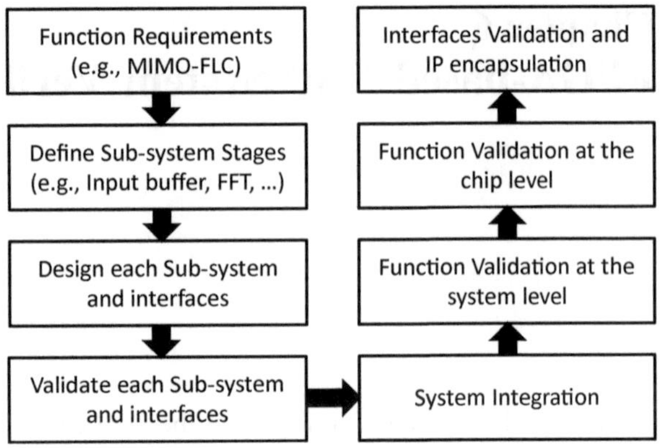

Fig. 6.1 FPGA-based digital system design and validation flow chart

technical point view, due to its advantages in the DSP algorithm design, validation and the eco-environment with FPGA, SoC and ASIC vendor tools.

The main task of the DSP system verification is to check two things: (1) during the normal operation situation, given the known inputs whether the outputs of the system are the ones expected; (2) during the extreme operation situation, given the boundary type inputs (which won't be seen during the normal operation, but theoretically they can be there) whether the outputs of the system can be reliably predicted and properly controlled. The latter scenario can be illustrated by a simple example of the FFT module utilized in the wireless LTE applications. Normalized data passing an FFT module will produce the frequency domain data with certain output magnitude, and in the fixed point FPGA-based environment, the output of the FFT needs to be properly scaled. If we could guarantee that only the normalized LTE-type frequency rich signal (normal operation) can reach to the input of FFT module, then we can scale the output of FFT module to maximize the resolution with limited fixed-point representation, while the same scaling factor is not propitiate (may cause overflow errors) if the normalized CW signal (extreme situation) appear at the input of the same FFT module, because a high-peak single frequency output will be produced according to the FFT operation mechanisms.

6.2.1 Simulink-based Pre-synthesize Functionality Verification Platform

Let's have a look at the first verification platform we'd like to introduce, i.e., a Simulink-based functionality verification platform, which is shown in Fig. 6.2, including the following four essential parts:

(1) Data Source: this module generates the testing data sequences or loads pre-generated or captured data sequences (e.g., LTE data streams) to stimulate the so-called reference model and designed DSP IP-core simultaneously.

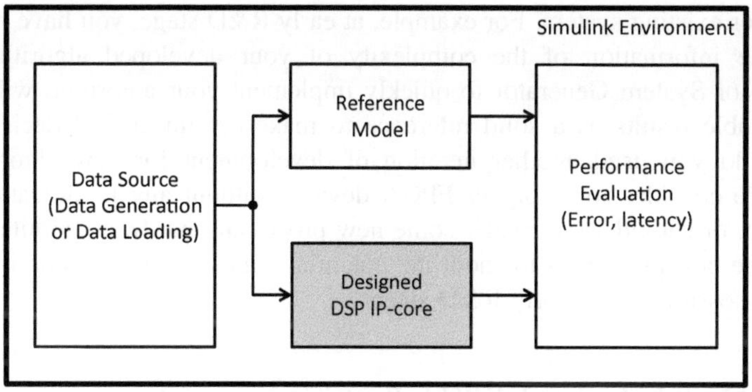

Fig. 6.2 Simplified diagram of Simulink-based IP-core functionality verification platform

(2) Reference Model: a double-precision floating-point based mathematical model should be in place as an ideal reference of the algorithm and it can be an existing Simulink model or a newly-designed but pre-verified Matlab model in a Simulink format.
(3) Designed DSP IP-core: this is the digital system designed by system generator [1] for Xilinx FPGA or Simulink with HDL coder toolbox [2] for generalized FPGA and ASIC.
(4) Performance Evaluation: this module will compare the fixed-point outputs of the designed DSP IP-core and double-precision floating-point outputs of the reference model module in terms of the relative/absolute errors and relative processing latencies.

Because the verification is performed in a kind of "software" environment, it provides a very flexible verification solution. For examples, using different random-generated or pre-characterized testing data sequences, we would verify the IP-core performance in both of the normal and extreme operation situations, respectively. And it will be quite easy to update and tune the designed DSP IP-core (both functionality and interfaces) to satisfy the system level requirement, significantly reducing the R&D time, which is very important at early research and prototyping stages for new technologies, like 5G wireless communication.

Though it is quite difficult for an algorithm engineer who doesn't have any knowledge of hardware to create a very efficient smart system (software and hardware co-operated unit), it is a trend that the software algorithm engineers are making more impacts on the whole system architecture in a project by the help of the tools provided by multiple EDA and semiconductor companies. The philosophy is simple, software and hardware co-optimized approach plays a very important role in order to solve certain problem with physically achievable solutions, an algorithm and architecture co-design approach is highly recommended.

Simulink (MathWorks) or System Generator (Xilinx) performs a unique role in quickly translating the mathematics-based algorithms into synthesizable HDL type hardware (FPGA or ASIC) instances, which can be tested separately and integrated in a bigger existing system. For example, at early R&D stage, you have no idea or very little information of the complexity of your developed algorithm, using Simulink or System Generator to quickly implement your algorithm will provide you valuable results as a solid reference to make certain critical decisions, like whether do you need another iteration of development for new algorithms, or should we consider some bigger FPGA devices without any modification of the algorithm, or should we consider some new processing hardware architecture and so on. The best practice is to shoot the potential issues as early as possible and as much as possible at the early R&D stages.

6.2.2 Matlab-Assisted FPGA Post-implementation Verification Platform

The effectiveness of the Simulink-based verification largely depends on the efficiency of the vendor tool, e.g., how efficient and how reliable to translate Simulink (or Matlab) models to certain HDL (e.g., VHDL, Verilog), which can be synthesized and implemented on a FPGA. And sometimes, you thought that the designed module is correct and producing outputs as expected, however when you synthesize and implement your designed IP-core with FPGA vendor tool (e.g., Xilinx Vivado/ISE, Intel Quartus II), your module will be "optimized" by the tools and then they may not be working in the way it is supposed to. No doubt, this is a mixed design and validation issue, which is suggested to be solved at design stage. The limitation is, we can only solve the issue we can see (or the problem that can be characterized somehow), in other words, we need to deal with the real hardware, i.e., FPGA chip, to see the input and output signals of designed IP-cores inside the chip. Certain tool, like Chipscope [3] or integrated logic analyzer (ILA) [4] is helpful to debug the signal inside the Xilinx FPGA, while for certain DSP IP-core, the functionality is not very straightforward to see by the signal itself, we need some other DSP IP-core validation approaches.

The second platform we'd like to introduce is a Matlab-assisted FPGA evaluation board based verification platform as shown in Fig. 6.3 including:

(1) A PC with Matlab: this part is utilized to generate data sources and produce reference model and its corresponding outputs. The same data source will be transmitted to the target FPGA evaluation board via certain interface, e.g., USB-to-UART (high speed interface, like PCI-express can be utilized at the cost of the more implementation and debugging efforts). And this part receives the data transmitted from FPGA as well, we then can carry out the performance evaluation in a MATLAB "software" environment by exploring the differences between the outputs from reference model and real designed DSP IP-core on a physical FPGA device.

(2) An FPGA evaluation board: for example, a Xilinx Virtex-7 evaluation board that is equipped with certain FPGA device, e.g., Virtex-7 FPGA and some connection interfaces, e.g., USB-to-UART convertor. The data source

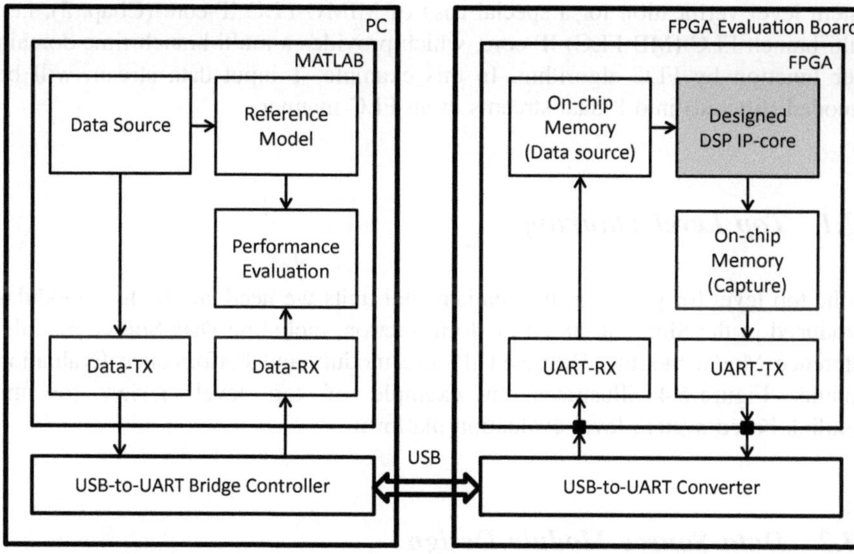

Fig. 6.3 Simplified diagram of Matlab-assisted FPGA evaluation board-based IP-core functionality verification platform

generated by MATLAB will be stored in the on-chip memory (e.g., Block RAMs [5] on the FPGA chip), and periodically stimulate the designed DSP IP-core. At the output, there is another group of on-chip memory to capture the IP-core outputs, one can design different capture scenario to thoroughly obtain the necessary output data sequences. Then those captured data sequences can be uploaded via USB-to-UART convertor to the PC side for analysis.

Using on-chip memory at the first place is highly recommended if the target FPGA chip has enough on-chip memory resources for your temporary validation, since the single-chip approach is relatively easy to handle and we avoid the potential issues introduced by dealing with external chip-to-chip interface, which is not the focus of the DSP IP-core validation. While, this simple SW and HW co-operated platform can be extended to large external memories data storage on the FPGA evaluation board if necessary by introducing certain memory interface controller and certain direct memory access (DMA) engine if some processors, e.g., ARM, MB or Nios II are involved in the targeting FPGA chip.

6.3 Verification at the System Level

The main task of the verification at the system level is to check the designed DSP IP-core functionality against the reference algorithm model in terms of the processing accuracy and processing latency. Let's have a look at an example of the system level verification for a special case of MIMO-FLC IP-core (Chap. 5), i.e., multi-branch FLC (MB-FLC) IP-core, which provides a multi-branch time domain filter function by FLC algorithm. In this example, 1 input data stream will be precoded (filtered) into 8 data streams in an FLC manner.

6.3.1 Top Level Thinking

At the top level for planning, the fundamental units we need are the four modules introduced in the Simulink-based platform section, including Data Source module, Reference Model module, Designed IP-core module and Performance Evaluation Module. Figure 6.4 illustrates an example of top level design for the Simulink-based system level evaluation platform.

6.3.2 Data Source Module Design

Data source module produces certain testing sequences to stimulate both the reference model and designed IP-core. Depending on the design strategy of the

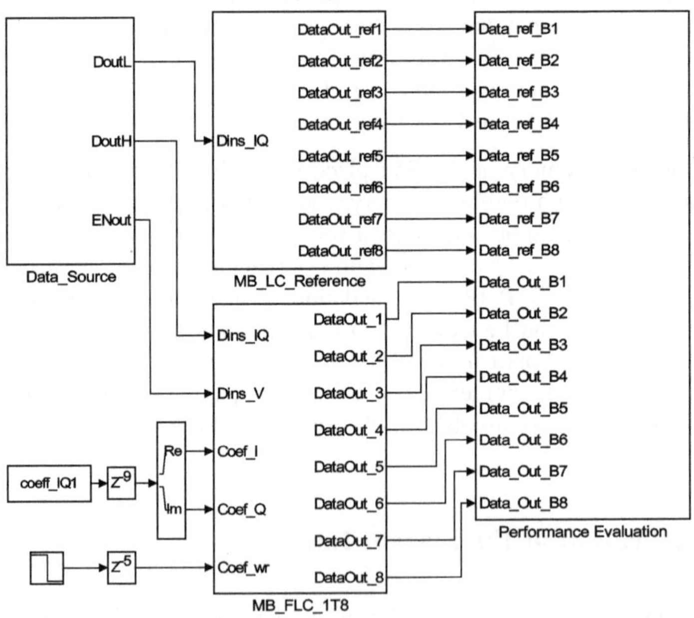

Fig. 6.4 An example of the Simulink-based system level evaluation platform at the top level

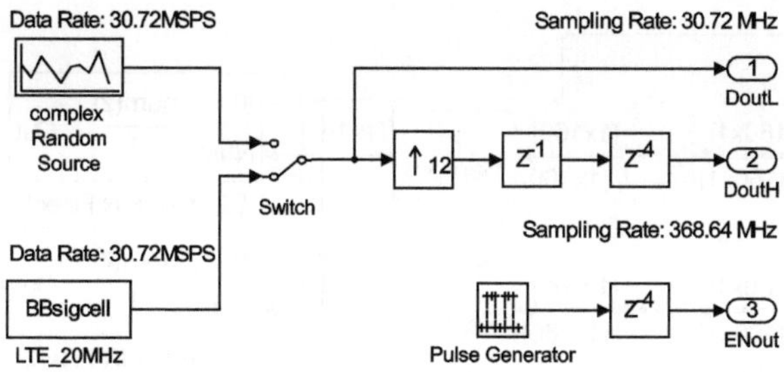

Fig. 6.5 An example of data source module, multiple sources with multi-rate outputs

IP-core, we may need multi-rate data sources as shown in Fig. 6.5, which produce the data sequences at the same data rate (e.g., 30.72 MSPS for 20 MHz LTE signal) but at different sampling rates, in the example, the sampling rate for reference model is the same as data rate (i.e., 30.72 MHz) while the sampling rate for the designed IP-core is 12 times higher, i.e., 368.64 MHz. And for robust validation, two types of data sources are usually utilized including the pseudo-random data source and delicate designed testing sequences (e.g., LTE signal for wireless applications).

6.3.3 Reference Model Module Design

Two strategies are commonly used to design a good reference model focusing on different perspectives. The first one is to create a corresponding reference sub-model for each sub-module in the designed IP-core, so that we can compare the designed IP and reference model at each intermediate processing stage. The other one strategy is to create a so-called cross-validation black-box type reference model using equivalent algorithm or architecture. The philosophy behind the latter strategy is to eliminate designer's human mistake by cross-validation, since following the strategy one may end up with "invalid" validation due to the same design mistake for both reference model and target IP-core. For example, fast linear convolution is a fast calculation algorithm to perform linear convolution, in order to validate the FLC IP-core, we could design a reference FLC module in the Simulink/MATLAB with every single processing stages involved, while alternatively, from function level point view, FIR filter carries out the same linear convolution behaviour, then we can create a reference MB-LC module in an FIR filter manner to avoid some "human-in-the-loop" design errors for FLC validation. Figure 6.6 illustrates the reference model we used for validating FLC module, i.e., 8-branch parallel FIR filters with 160 complex coefficients for each branch.

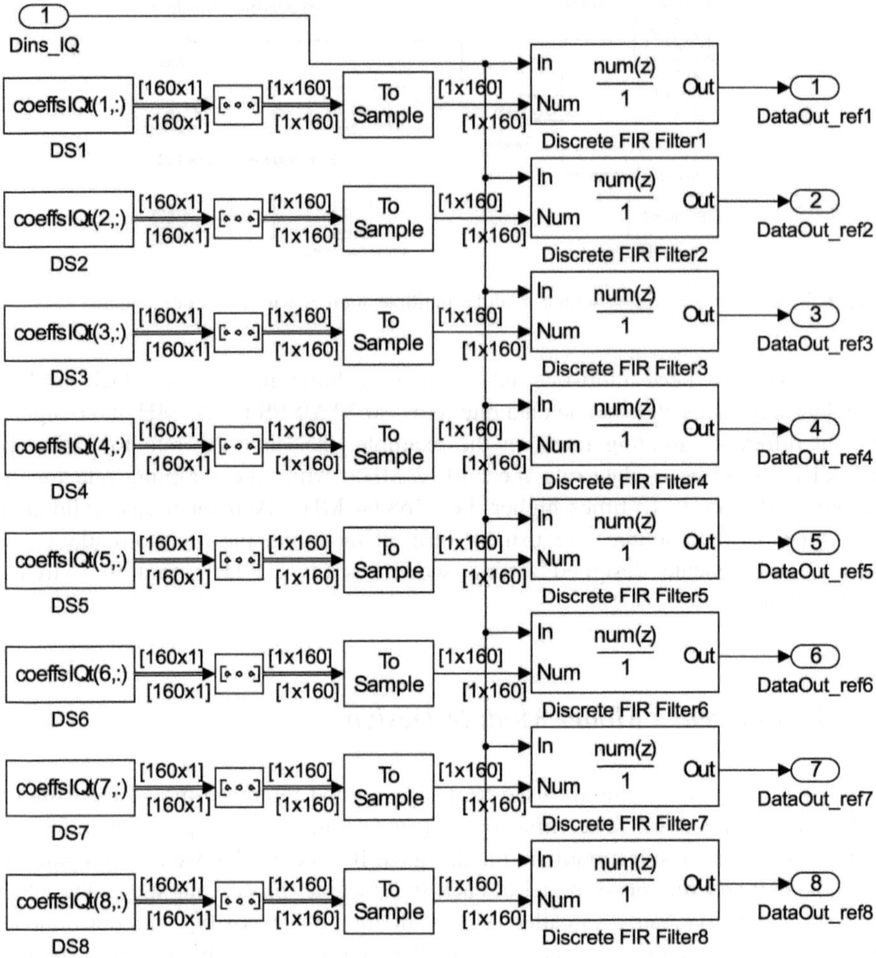

Fig. 6.6 Reference model design using equivalent FIR architecture for FLC validation

6.3.4 DSP IP-Core Module Design

Figure 6.7 illustrates the designed MB-FLC IP-core using Vivado System Generator tool, as a special case of MIMO-FLC IP-core (Chap. 5), MB-FLC IP-core applies multiple coefficient sets for one input data steam and produces multiple precoded outputs. Detailed FLC algorithm is available in Chap. 5. Particularly, for this example, we implement a 1-to-8 MB-FLC core using 512-size FFT/IFFT.

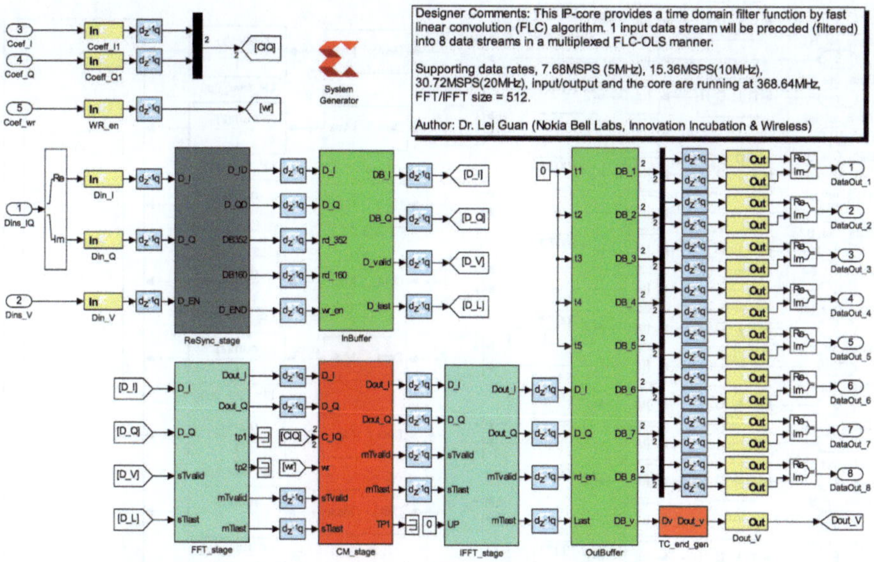

Fig. 6.7 Designed MB-FLC IP-core using system generator

6.3.5 *Performance Evaluation Module Design*

For performance validation, the same LTE 20 MHz test mode signal (64 K samples) was used to stimulate both of the reference MB-LC module and MB-FLC core, and we compared the outputs of fixed-point FPGA MB-FLC IP-core (designed using System Generator) with the outputs of floating-point MB-LC reference model. Specifically, the reference FIR-based MB-LC module in the Simulink utilizes 8 different sets of 160-tap floating-point complex coefficients, which are converted to 8 sets of fixed-point (32-bit for complex I and Q coefficients) FLC coefficients in the MATLAB, then the MB-FLC IP-core loads those coefficients (512 complex numbers per set, and 4096 complex numbers in total for 8-branch FLC core) and performed MB-FLC function.

In the performance evaluation module as shown in Fig. 6.8, we calculate the absolute errors in the magnitude between time-aligned MB-LC outputs and MB-FLC outputs for all 8 branches. For all of the 8-branch, less than $\pm 0.04\%$ error were achieved with 16-bit resolution (16-bit I signal and 16-bit Q signal) for the MB-FLC outputs (Fix_16_14) as shown in Fig. 6.9. If lower error rate needs to be achieved, higher resolution design can be used at the cost of extra resources and processing latencies.

In the performance evaluation module, processing latency performance needs to be assessed as well, as shown in Fig. 6.8, the delay blocks along the reference outputs (e.g., 10,073 in this example), are in place due to the processing latency introduced by the designed MB-FLC IP-core. And taking the equivalent processing

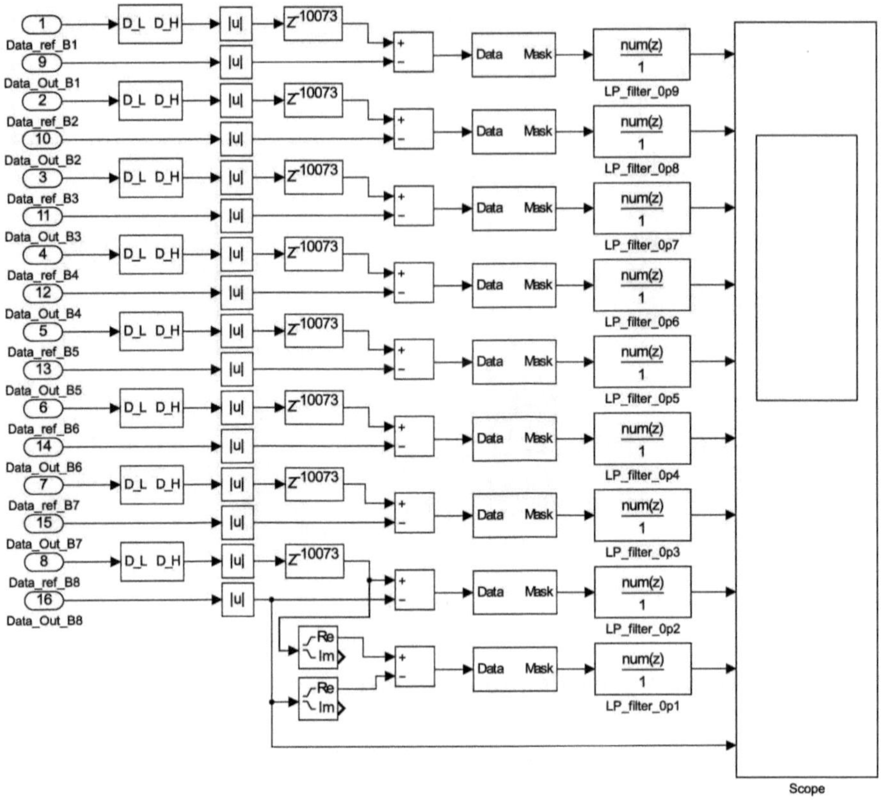

Fig. 6.8 Performance evaluation module

clock, 368.64 MHz into consideration, 10,073 cycles of 368.64 MHz clock is roughly 27 μs, which means that the processing latency from the first input sample arriving at the MB-FLC IP-core input port to the first available sample at the output port is 10,073 clock cycles, i.e., 27 μs. Please note this 27 μs processing latency will be the real physical processing latency in the FPGA, the actual software simulation duration in the Simulink environment will be generally much longer than the physical hardware processing duration in the FPGA.

6.4 Verification at the Chip Level

After function validation, next task is to verify the performance of the designed IP-core on the real FPGA chip regarding actual resource utilization, timing performance and power estimation. In order to evaluate the silicon performance, we

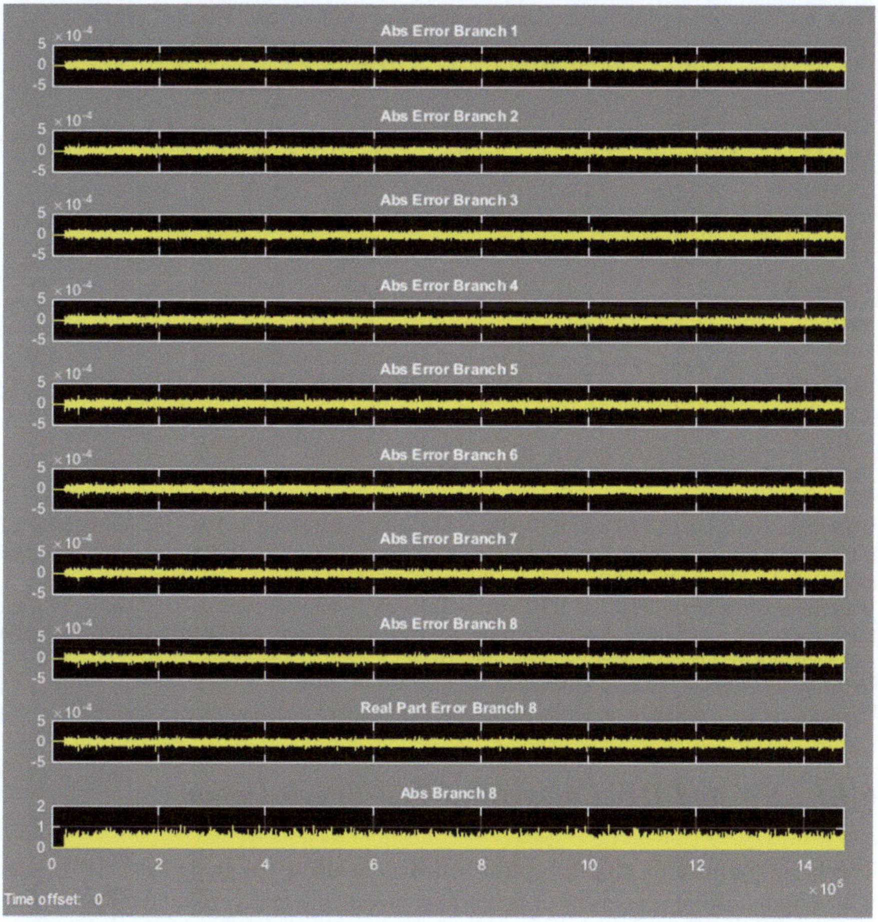

Fig. 6.9 Performance evaluation illustration by the absolute errors of the magnitude

need to synthesize and implement the model-based DSP IP-core, e.g., MB-FLC IP-core, using some FPGA vendor tools, e.g., Vivado/ISE design tool for Xilinx FPGAs, Quartus II for Intel FPGAs. For example, we synthesize and implement the MB-FLC IP-core in the Xilinx Vivado 2015.4 environment targeting Kintex Ultrascale device, with full hierarchy in synthesize setting and default implementation strategy settings. For solution effectiveness comparison, sometime we synthesize and implement a few alternative solutions. For example, Table 6.1 illustrates the resource utilization, latency and power consumption comparison between compact semi-parallel FIR-based linear convolution solution [6] and alternative MB-FLC solution with different configurations [7].

Table 6.1 Chip-level performance comparison between MB-LC and MB-FLC (8-branch)

	Total[1]	LC-core[2]		FLC-core[3]		FLC-core[4]	
		Usage[5]	Percent	Usage	Percent	Usage	Percent
DSP48s	5520	240	4.35	49	0.89	129	2.34
BRAMs	2160	2	0.09	27	1.25	27	1.25
LUTRAM	293,760	1360	0.46	5496	1.87	5451	1.86
LUT	663,360	8824	1.33	11,850	1.79	9341	1.41
FF	1,326,720	38,489	2.90	18.395	1.39	13,756	1.04
Power (W)[6]	–	4.41		1.47		1.45	
Latency (μs)	–	~0.23		~27.32		~27.32	
Data rate	–	30.72 MSPS		30.72 MSPS		30.72 MSPS	
Core clock (MHz)	–	491.52		368.64		368.64	

[1]Based on the Kintex Ultrascale XCKU 115 device
[2]Complex linear convolution with semi-parallel FIR architecture with 160 coefficients each
[3]Fast linear convolution, FFT and IFFT size = 512 with 16-bit phase and data resolution, FFT and IFFT cores are configured for minimizing DSP48 utilization, 3-DSP48 complex multiplier architecture with non-DSP48 based butterfly unit for FFT
[4]Fast linear convolution, FFT and IFFT size = 512 with 16-bit phase and data resolution, FFT and IFFT cores are configured for minimizing LUT/FF utilization, 4-DSP48 complex multiplier architecture with DSP48 based butterfly unit for FFT
[5]Post-implementation resource utilization
[6]Dynamic power consumption estimated by Vivado tool, not including device static power

6.4.1 SW and HW Co-operated Test-Bench Design

The basic purpose of chip-level test-bench is to drive the designed DSP IP-core with certain testing sequences or conditions in the chip and verify the outputs of the IP-core are the ones expected. Before starting test-bench design, requirement analysis should be carried out focusing on the three essential aspects below:

(1) System Functionality. The main function of the test-bench platform is to provide a human to machine interactive system, which will be utilized to verify the ideas (software algorithms) on a machine carrier (hardware) for solving a particular engineering or research problem. The essential building blocks required can be generally grouped into two categories, software (SW) and hardware (HW), as illustrates in Table 6.2. In this book, we focus on the scenario that is using MATLAB as algorithm SW host and FPGA as algorithm HW implementation host.
(2) System Flexibility. For validating the DSP algorithms at prototyping stage, the basic requirement is to create a platform with flexible pattern generator function and data acquisition function. A general purpose PC with MATLAB, an ideal scientific SW environment, is chosen to provide a relevant high DSP algorithm developing flexibility in SW; while a FPGA evaluation board is

Table 6.2 Fundamental function requirement list

Location	Functions	Examples
SW (MATLAB on a PC)	Graphic user interface	Control panel
	User defined applications	Data source generator, DPD coefficient and MIMO-FLC coefficients extraction
	Data transmission interface (towards FPGA)	UART, Ethernet, USB, PCI-E
HW (FPGA)	Data transmission interface (towards PC)	UART, Ethernet, USB, PCI-E
	Data storage	BRAMs, DDR RAMs
	Clock management	MMCM, clock distribution
	User designed IP-cores	Digital predistorter and MIMO-FLC -based precoding
	Data transmission interface (towards next stage)	LVDS, JESD204B, CPRI

used to provide a physical foundation for the re-configurable hardware system design. We would like to emphasize that, once the verification stage has been done, it is prudent to migrate the design from discrete components-based platform to a more integrated single chip solution, e.g., Xilinx Zynq SoC device or Intel Stratix 10 device, for maximizing the overall system performance of your ideas (algorithms).

(3) System Expandability. For achieving a robust universal test-bench platform solution, the system functionality and flexibility are the basic requirement, while the system expandability will lay a far-seeing constraint to design an industrial integrated platform for now and the future. Targeting a platform with minimum resource cost, only the fundamental functions will come into our consideration. Those functional digital blocks are independently designed in logic level so that they can be easily upgraded and expanded. For example, for simplifying the interacting between PC and FPGA, the data communications interface between PC and FPGA evaluation board is based on the Universal Asynchronous Receiver Transmitter (UART) protocol. Upgrading this UART interface to other protocol will only affect the interface logic, and the other logic, such as the designed IP-core will stay the same.

6.4.2 Test-Bench SW Design in MATLAB

As shown in Table 6.2, SW category contains three fundamental functions: data transmission, user defined application and graphic user interface (GUI), which can be efficiently designed in three separate functional layers as shown in Fig. 6.10.

The fundamental purpose of L1 is to manage the data transmission between upper layer (L2) SW and the lower layer HW. A typical L1 can be realized as

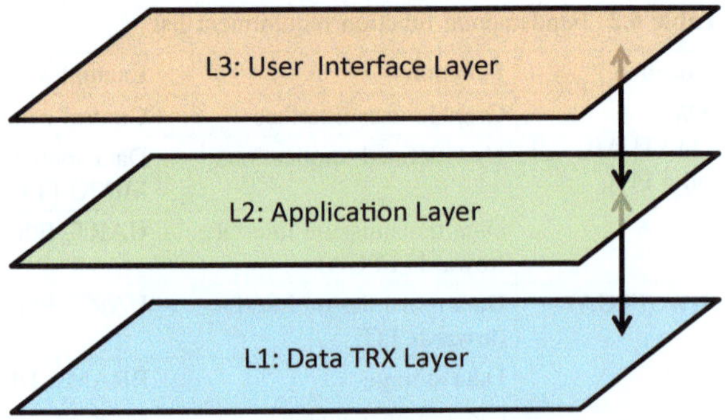

Fig. 6.10 High level SW design in MATLAB

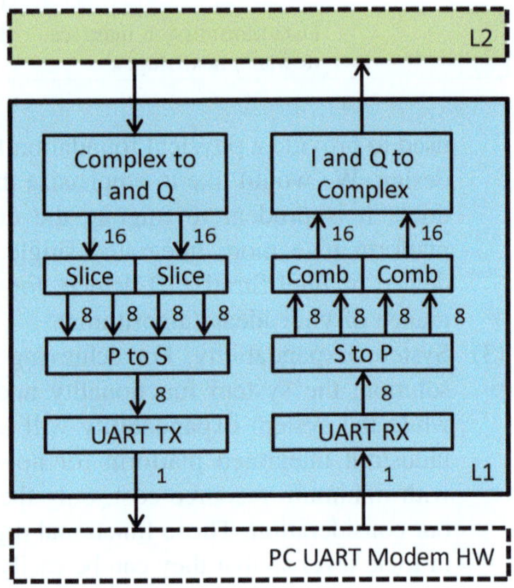

Fig. 6.11 A typical design of the L1 SW in MATLAB

illustrated in Fig. 6.11, including a UART transmitter and receiver, a data convertor between serial and parallel, data combiners and slicers, and data convertors between real number I and Q signal and complex number signal.

L2 functions typically include data generator, user algorithms (e.g., DPD, MIMO-FLC) and performance evaluation, which are the core functions of the SW in the MATLAB. Figure 6.12 illustrates a typical L2 design in MATLAB SW environment. Both data transmission and control instruction transmission should be properly managed to smooth the executions of the SW.

On top of the L2, L3 provides a graphic user interface enabling a good experience when dealing with algorithm validation by the other designers/researcher/engineers in the team. Only the system level functions are necessary to be visible in the L3 for system level functionality validation of the DSP algorithms. Figure 6.13 shows an example of L3 GUI designed for DPD applications [8, 9] and Fig. 6.14

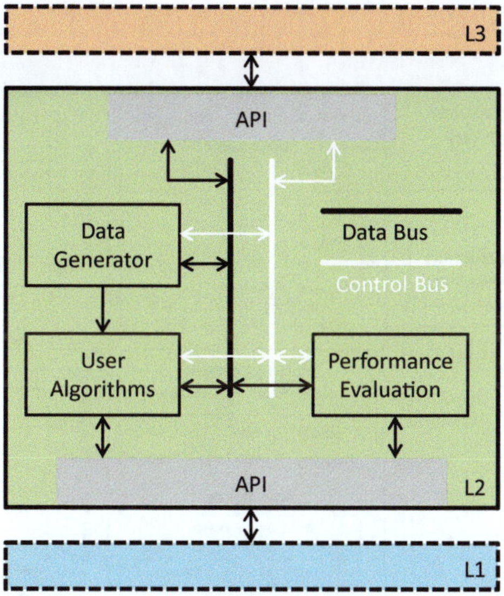

Fig. 6.12 A typical design of the L2 SW in MATLAB

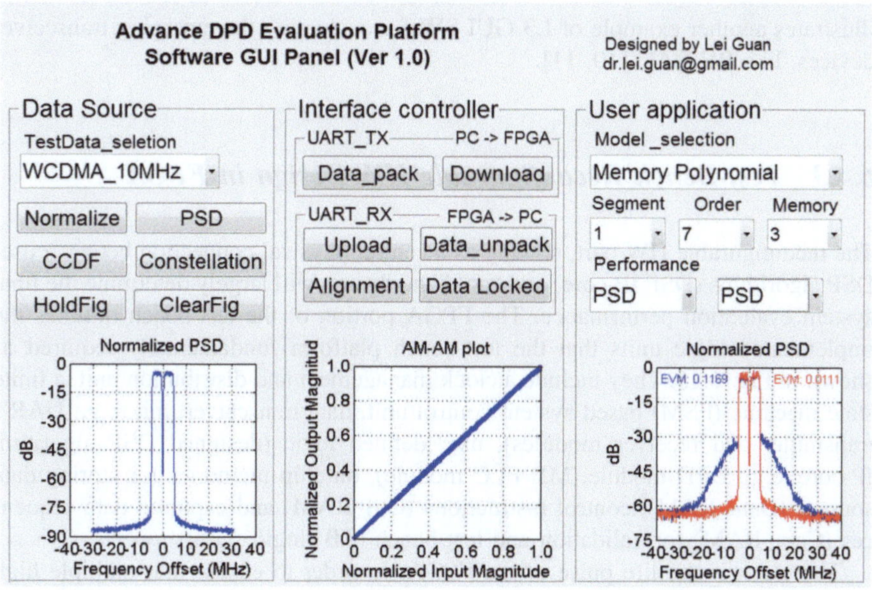

Fig. 6.13 An example of the L3 GUI SW in MATLAB for DPD evaluation platform

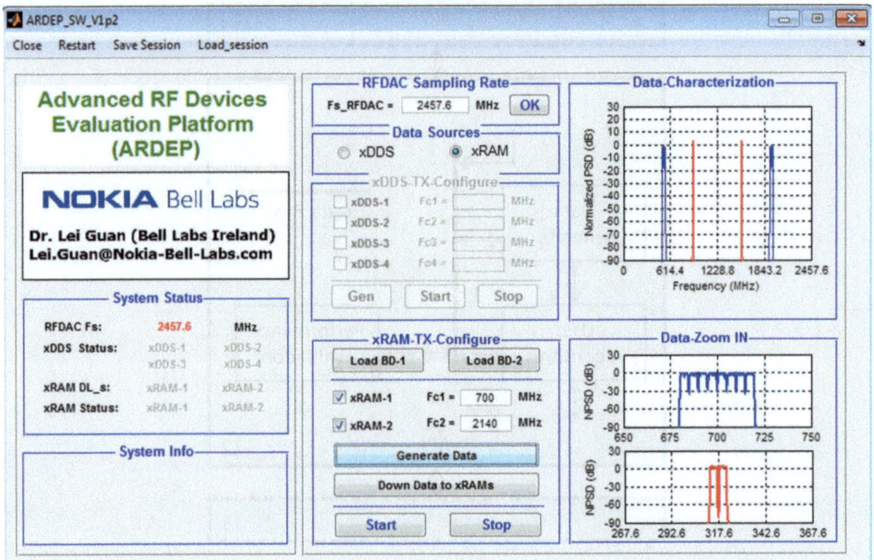

Fig. 6.14 An example of L3 GUI SW in MATLAB for RF-DAC evaluation platform

illustrates another example of L3 GUI SW for evaluating the emerging transceiver devices, like RF-DAC [10, 11].

6.4.3 Test-Bench Reconfigurable HW Design in FPGA

The reconfigurable HW part, serving as a central physical connection between user DSP algorithms (DSP IP-core) and "cold" hardware, will largely determine the final system evaluation performance. The FPGA portion of the test-bench outlines the implemented logic units that the test-bench platform fundamentally required as shown in Fig. 6.15. They include a clock management and distribution unit, a finite state machine (FSM) based system control unit, data transceiver unit (e.g., UART transmitter and receiver modules), user defined logic (designed DSP algorithm IP-core, e.g., DPD module, MB-FLC module), on-chip memories for storing data sources (Data_RAM), control instructions (Ctrl_RAM) and captured data sequences (Capt_RAM) for validation and test-bench (TB) logic.

The clock is the life pulse of the FPGA. In order to ensure that multiple high speed clocks (CLK_SYS and CLK_IP) are properly distributed across required FPGA banks, we chose vendor designed IP-core, like Xilinx clock management (MMCM) core, which provides an easy, interactive way of defining and specifying clocks. And for some low speed clocks, e.g., the one used for UART transceiver (CLK_UART at 921.6 kHz), we use a simple counter based clock distribution (or

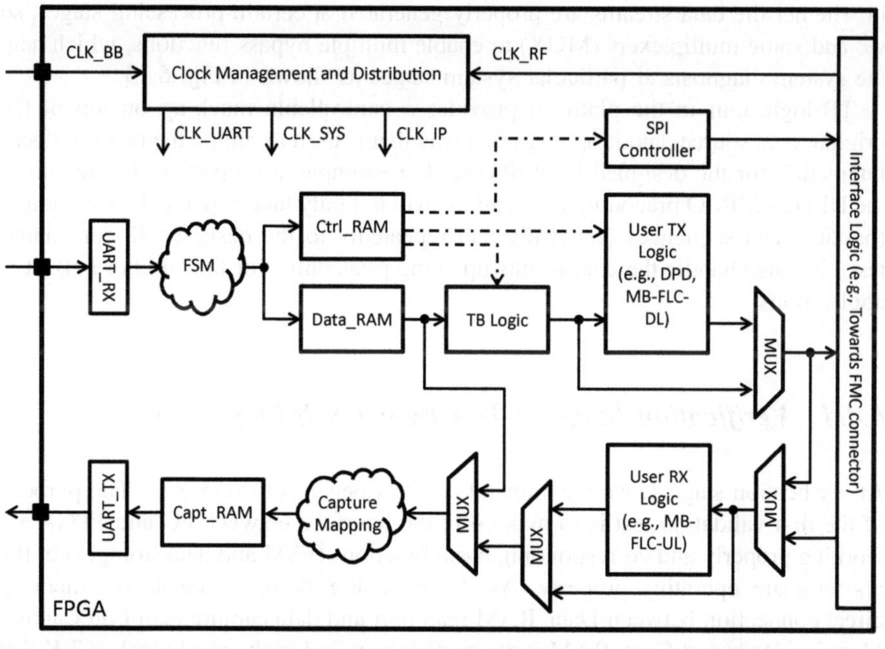

Fig. 6.15 An example of simplified diagram of FPGA architecture for the test-bench design

frequency divider) to provide the required clock rate. In order to achieve synchronous data transmission between digital hardware (e.g., FPGA evaluation board) and next stage analog front-end in the wireless applications, a high-speed clock should be generated from a stable clock source (CLK_RF) and it will be used to drive both digital hardware (FPGA) and corresponding data convertors (e.g., DAC and ADC) on the analog front-end. While since the data transmission between PC and digital hardware does not require synchronize transmission, those logics can be driven by the on-board oscillator (CLK_BB).

As mentioned before, high-performance on-chip RAMs are always the first choice for creating small to medium-size user storage. For example, in this platform, we utilized Xilinx's Block Memory Generator [5] to build up three independent RAMs for control purpose, data TX and RX purposes. Particularly the data RAMs (including Data_RAM and Capt_RAM) are built based on dual-port RAM. For example, at data transmitter path, Data_RAM receives the data from PC at slow rate (CLK_UART, e.g., 921.6 kHz) and stimulates the user logic at required high rate (CLK_IP, e.g., 368.64 MHz), similar design configurations are used for the data receiver path, i.e., one port of the Capt_RAM is driven by high-speed clock for properly synchronized data capture, and the other one port uploads the captured data to the PC in an on-demand way at slow speed clock.

For creating a robust test-bench, it is crucial to reserve certain flexibility by inserting some test points in the system. At the system level, we are more interested in whether the data streams are properly generated at certain processing stages, so we add some multiplexers (MUX) to enable multiple bypass functions, which help the system diagnosis at particular system stages as shown in Fig. 6.15.

TB-logic unit in the platform provides a controllable mock-up on top of the original data source, and it operates in a way that is used to mimic the real interfaces interaction for the designed DSP IP-core. For example, a proper TB_Logic unit in the DPD or MIMO precoding applications will not only take care the data streaming (passing data sequences constantly and repeatedly to the designed IP-core under test), but also handle the coefficients updating procedures for the DPD or MB-FLC applications.

6.4.4 Verification Stage-1: Test-Bench Self-loop Tests

At verification stage 1, we recommend two test-bench self-loop tests. The purpose of the first validation test is to make sure the interface between PC and FPGA are working properly and corresponding basic functions (FSM and data storages) of the platform are operating properly. As shown in Fig. 6.16, we create a temporary direct connection between Data_RAM read port and data capture path (via Capture Mapping logic and Capt_RAM write port) at required high speed clock (CLK_IP)

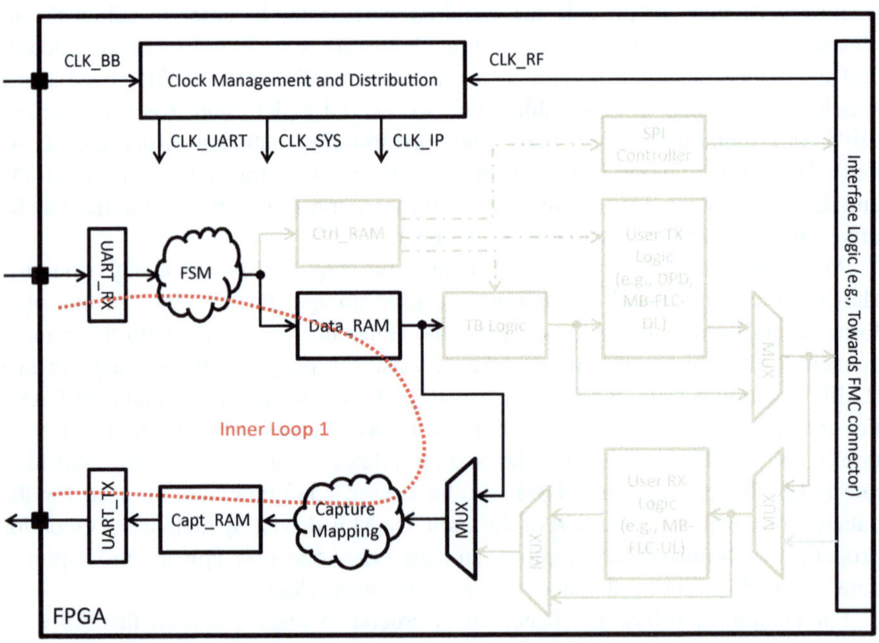

Fig. 6.16 Test-bench self-loop test (inner loop 1 validation)

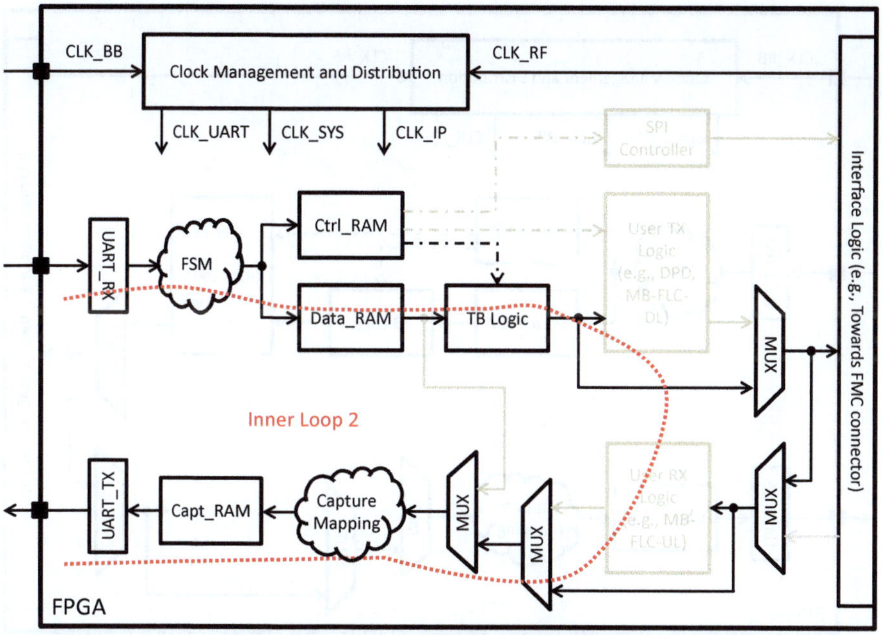

Fig. 6.17 Test-bench self-loop test (inner loop 2 validation)

domain in the FPGA. We generate a testing complex baseband data sequence in MATLAB, convert the data into bytes and send them to the FPGA-eva board. Then we capture the data and upload them to the MATLAB to compare with the original complex baseband data sequence. A valid test will ensure that the original data and captured data should be identical with known latency.

After validation of the inner loop 1, the second self-loop validation is to verify the functionality of the TB_Logic unit (together with control function in the Ctrl_RAM), which is an essential module to ensure that the user logic has been properly stimulated. As shown in Fig. 6.17, we perform an inner loop 2 self-loop testing by enabling several multiplexers in the platform.

6.4.5 Verification Stage-2: Designed IP-Core in-the-Loop Test

Next, at validation stage 2, we activate the designed IP-core in the loop as shown in Fig. 6.18 to evaluate its performance in terms of functionality, resource utilization, power consumption and timing performance. In order to avoid messed-up situation between designed IP-core and other logical units of the test-bench inside the FPGA, the best practice is to add extra pipeline registers between designed IP-core and

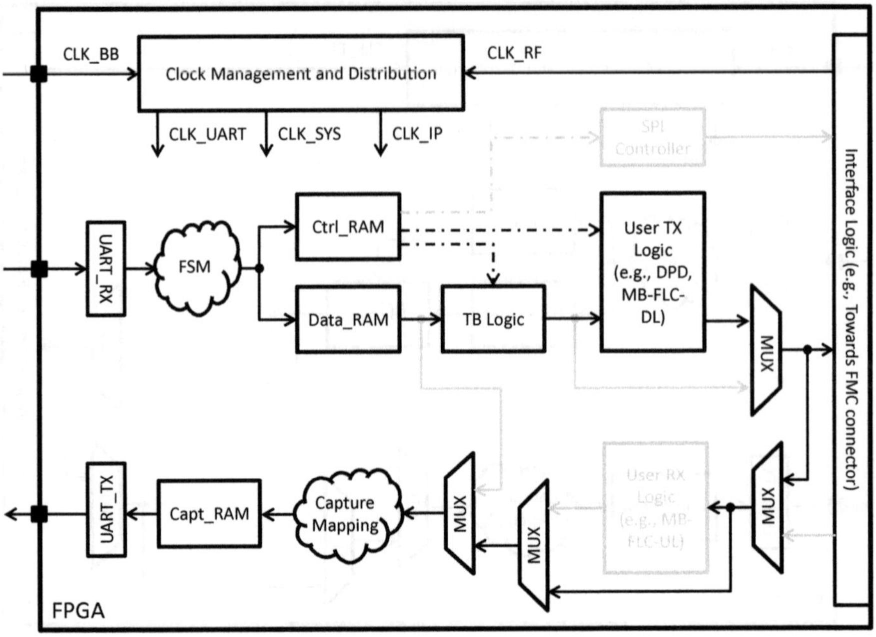

Fig. 6.18 Designed IP-core in-the-loop test (user logic at the TX path is used as an example)

other logical units, so that the designed IP-core can be logically isolated providing better timing performance and easier timing analysis break points. By defining different capture mapping scenarios, we create varied event-driven data capture scenarios, for example the outputs of the MB-FLC contain multiple complex data streams from different branches, by using certain capture mapping, we can capture the data in any one of the output branches.

One of the most important purposes of this test is to ensure the designed IP-core is operating as expected on the real FPGA chip given proper inputs. Comparing to the reference model in the MATLAB, we should only observe certain quantization errors which are associated with fixed-point design and FPGA processing if the designed IP-core has been designed properly. And actually, we should observe some performance results similar as the one shown in Fig. 6.9.

6.4.6 Verification Stage-3: Whole System in-the-Loop Test

Next verification stage is to evaluate the performance of the designed DSP IP-core with the whole system in-the-loop, i.e., the real physical application oriented tests. In the first example as shown in Fig. 6.19, we use the SW and HW co-operated test-bench platform to evaluate the nonlinear convolution based DPD for PA

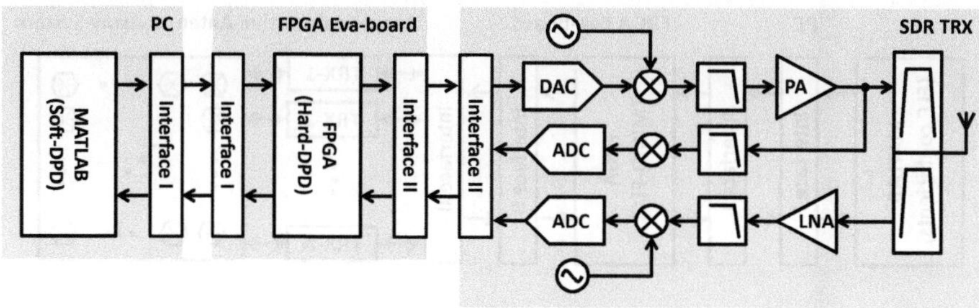

Fig. 6.19 Whole system in-the-loop test for DPD application

linearization in the software defined radio (SDR) wireless transceiver [12–17]. Two operation modes (Soft-DPD and Hard-DPD) are designed to accelerate the algorithm development and validation.

In the so-called soft-DPD mode, the DPD signal processing is done mainly in the MATLAB software environment. For instance, we first generate a testing sequence, e.g., LTE data stream, and then download the data stream to the Data_RAM in the FPGA to stimulate the RF chain. Secondly, we capture the feedback signal coupled from nonlinear PA output and perform model extraction in the MATLAB software environment. Thirdly, we predistort the signal using extracted coefficients in a nonlinear convolution way in the MATLAB with certain power control (to maintain the same power level as the one without DPD function). Finally, we download the predistorted data stream into the Data_RAM in the FPGA (bypassing the user logic in the FPGA) to combat for the nonlinearity introduced by the RF PA under test.

In the so-called hard-DPD mode, the main nonlinear convolution -based predistorter (see Chap. 4) is in place as user defined logic in the FPGA, we estimate the DPD coefficients in the MATLAB environment or in the on-chip processor (e.g., Microblaze or ARM) environment. Only the coefficients are required to transfer from other blocks to the FPGA DPD IP-core. In this mode, the signal processing part has been done mainly in the FPGA hardware environment, which performs robust experimental algorithm validation.

In the second example as shown in Fig. 6.20, we use the SW and HW co-operated test-bench platform to verify the essential signal processing part of the massive MIMO wireless system [18, 19]. Like DPD approach, the main processing blocks required by massive MIMO system can be divided into two main groups, real-time function and non-real-time function. The real-time processing function includes the precoding at downlink and the combining at uplink, while the non-real-time processing functions include channel estimation, precoding and combining weights calculation. Simply speaking, the MIMO processing can be generalized into two parts: real-time convolution-based precoding/combining (or frequency domain multiplication-based precoding/combining) and non-real-time precoding/combining coefficients extraction (including channel estimation and

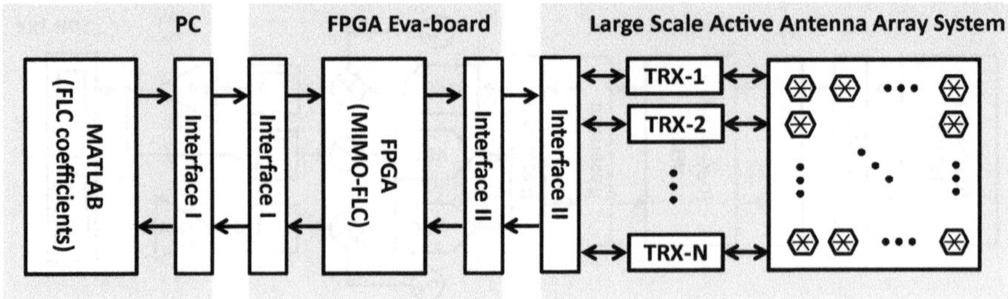

Fig. 6.20 Whole system in-the-loop test for generalized MIMO application

weighting coefficients calculation). Depending on the requirement of processing latency, some other SW hosts, e.g., general purpose processor (GPP), or graphic processing unit (GPU) can be utilized to replace MATLAB as well.

6.5 Conclusions

In this chapter, we outlined and explained the best practices for validating the DSP IP-core designed by the model-base approaches. Two useful platforms were introduced to provide robust and fast functionality validation at both of the system level and FPGA chip level. We discussed the architecture, the system design strategies and detailed verification steps of the validation platforms, which turn out to be cost-effective and flexible validation solution for fast FPGA-based DSP algorithms, e.g., nonlinear convolution based DPD functions, linear convolution based massive MIMO precoding functions. The SW and HW co-operated platforms presented in this chapter can also be utilized to design and evaluate the emerging wireless applications, such as full-duplex transceiver concept, compact single-chip massive MIMO digital solution, LiFi wireless transceivers and so on.

References

1. Xilinx (2015) Model-based DSP design using system generator. User Guide, UG897
2. MathWorks (2016) HDL coder user's guide. R2016b
3. Xilinx (2012) ChipScope pro software and cores. User Guide, UG029
4. Xilinx (2016) Vivado design suite tutorial, programming and debugging. UG936
5. Xilinx (2016) Block memory generator v8.3. LogiCORE IP product guide, PG058
6. Guan L (2016) Xilinx FPGA enables scalable MIMO precoding core. Excellence Wirel Commun Xilinx Xcell J, issue 94
7. Guan L (2015) Compact scalable frequency dependent precoding and its FPGA implementation for 5G massive MIMO wireless systems. IET Electron Lett 51(23):1937–1939
8. Guan L, Kearney R, Yu C, Zhu A (2013) High-performance digital predistortion test platform development for wideband RF power amplifiers. Int J Microw Wirel Technol 5(2):149–162

9. Guan L, Zhu A (2011) Dual-loop model extraction for digital predistortion of wideband RF power amplifiers. IEEE Microw Wirel Compon Lett 21(9):501–503
10. Guan L (2015) Evaluating the linearity of RF-DAC multi-band transmitters. Excellence Wirel Commun Xilinx Xcell J, issue 91
11. Guan L, Rulikowski P, Kearney R (2016) Flexible practical multi-band large scale antenna system architecture for 5G wireless networks. IET Electron Lett 52(11):970–972
12. Guan L (2012) High performance digital predistortion for wideband RF Power Amplifiers. Ph.D. thesis, University College Dublin
13. Guan L, Zhu A (2014) Green communications: digital predistortion for wideband RF power amplifiers. IEEE Microw Mag 15(07):84–99
14. Kelly N, Zhu A (2016) Low complexity stochastic optimization-based model extraction for digital predistortion of RF power amplifiers. IEEE Trans Microw Theory Tech 64(05):1373–1382
15. Yu C, Cao W, Guo Y, Zhu A (2015) Digital compensation for transmitter leakage in non-contiguous carrier aggregation applications with FPGA implementation. IEEE Trans Microw Theory Tech 63(12):4306–4318
16. Guan L, Zhu A (2012) Optimized low-complexity implementation of least squares-based model extraction for digital predistortion of RF power amplifiers. IEEE Trans Microw Theory Tech 60(3):594–603
17. Yu C, Guan L, Zhu E, Zhu A (2012) Band-limited Volterra series-based digital predistortion for wideband RF Power amplifiers. IEEE Trans Microw Theory Tech 60(12):4198–4208
18. Weldon MK (2016) The future X network: a Bell Labs perspective. CRC Press, New York
19. Guan L, Giordano LG, Bonfante A (2016) A flexible HW and SW co-operated baseband research platform for massive MIMO system. Bell Labs internal technical report

MIX
Papier aus verantwortungsvollen Quellen
Paper from responsible sources
FSC® C105338

If you have any concerns about our products,
you can contact us on
ProductSafety@springernature.com

In case Publisher is established outside the EU,
the EU authorized representative is:
Springer Nature Customer Service Center GmbH
Europaplatz 3, 69115 Heidelberg, Germany

Printed by Libri Plureos GmbH
in Hamburg, Germany